ME MEDICINE vs. WE MEDICINE

ME MEDICINE vs. WE MEDICINE

Reclaiming Biotechnology for the Common Good

DONNA DICKENSON

COLUMBIA UNIVERSITY PRESS   NEW YORK

COLUMBIA UNIVERSITY PRESS
PUBLISHERS SINCE 1893
NEW YORK   CHICHESTER, WEST SUSSEX
cup.columbia.edu

Library of Congress Cataloging-in-Publication Data
Dickenson, Donna.
  Me medicine vs. we medicine : reclaiming biotechnology for the common good / Donna
Dickenson.
      p. ; cm.
  Includes bibliographical refernces and index.
  ISBN 978-0-231-15974-6 (cloth : alk. paper)—ISBN 978-0-231-53441-3 (e-book)
  I. Title.
  [DNLM: 1. Individualized Medicine—ethics.   2. Biotechnology—ethics.   3. Public Health
Practice—ethics.   4. Social Justice. WB 102]
  174.2—dc23

                                                                                          2012037582

29.95

Columbia University Press books are printed on permanent and durable acid-free paper.

This book is printed on paper with recycled content.
Printed in the United States of America
c 10 9 8 7 6 5 4 3 2 1

JACKEY DESIGN: CHANG JAE LEE

References to websites (URLs) were accurate at the time of writing. Neither the author nor
Columbia University Press is responsible for URLs that may have expired or changed since the
manuscript was prepared.

CONTENTS

PREFACE

MY FATHER DIED OF HODGKIN'S DISEASE WHEN HE WAS TWENTY-FIVE. I remember visiting him, encased by a frightening iron lung, in a Veterans Administration hospital near New York. (The next day he wrote a letter reassuring me, with a cartoon of his feet sticking comically out of the iron lung.) The VA accepted responsibility for his medical treatment, and later for my college education, because he'd contracted the disease during his naval service in World War II. Of course, to me, at the age of four, a hospital was just a hospital, whether public or private—or more likely, just a strange and scary place. It was only later that I became aware that the VA hospitals where my father and grandfather underwent treatment were something to be very grateful for: a widespread and accessible example of publicly funded care, one form of what I call "We Medicine."

In this book, I examine why We Medicine has instead become increasingly distrusted—why public health has often come to be seen as a public enemy—and why its opposite number, personalized healthcare—what I call "Me Medicine"—is gaining the ascendancy. Even in the increasingly individualized American medical system, advocates of personalized medicine claim that healthcare isn't individualized enough. With the additional glamour of new biotechnologies such as genetic testing behind it, Me

Medicine appears to its advocates as the inevitable and desirable way of the future.

Despite my early experiences, I don't automatically assume that Me Medicine is bad and We Medicine is good, even though the proponents of personalized healthcare very rarely challenge their own preconception that the reverse is true. Instead, I do my level best to give a balanced, evidence-based account. Given that the scientific evidence doesn't dictate that you have to be ready to accept the supposed revolution of personalized medicine, where do we go from here? Is Me Medicine or We Medicine the way forward? Must we choose?

The answer to this question can't just depend on the scientific evidence, because moral and political choices are involved—although it is important to know what that evidence is before we can give an informed answer. That's a very major chunk of what I try to do in this book: to give you the evidence about the various forms of personalized medicine, so that you can decide for yourself.

But in addition to offering a reality check for personalized medicine, I also try to do something more: to ask some big questions about the role of the communal and the individual in healthcare and in our civic life. I'm a philosopher by training, so I know how far back these debates go and how they go on and on. They're embodied and emblematized today in the personalized healthcare movement, but they have a much older pedigree. Part of the task of this book is to set personalized healthcare in that broader social and political context. Whatever explains the rise of Me Medicine, it isn't just the science behind it. Likewise, the causes of popular rebellion against forms of We Medicine such as vaccination aren't rooted only in medical evidence.

If the scientific evidence alone doesn't explain the rise of Me Medicine and the comparative decline of We Medicine, what does? I look critically at four possible factors, some of which will turn out to be more convincing than others: threat, narcissism, corporate interests, and the dominance of autonomy and choice in our thinking. And I ask this crucial question: how did we move from what was originally presented as a communitarian vision for the new genetic biomedicine to the now-dominant personalized medicine paradigm?

Throughout the book, I'm driven by the question of how we can reclaim biotechnology for the common good. In the final chapter, the concept of the commons will be a major focus. Reclaiming biotechnology for the greater good will involve resurrecting the commons. Although some attention has focused on the genome as the common property of humanity, many other aspects of modern biomedicine could, and I think should, be considered a commons. For example, when enough children are vaccinated against diseases such as measles, the resulting population, or "herd," immunity also benefits neonates, the elderly, and others too frail to be vaccinated. But when too many parents opt out of vaccination, population immunity is diminished, just as overgrazing or overfishing subtracts from a common resource in land or fisheries. In the extreme, the common resource is put fatally at risk.

I've been interested in the concept of the commons in biomedicine for some time, primarily in terms of the commodification of the body. Much of my work is devoted to showing how concepts from the law of property, such as the commons, can be applied to medicine in such a way as to protect the rights of the vulnerable. In my academic books *Property, Women, and Politics* (1997) and *Property in the Body* (2007), as well as in my popular-science book *Body Shopping* (2008), I examine the way in which the human body, particularly women's tissue, is becoming an object of commercial exploitation, but I also offer strategies using property concepts to overcome that sort of exploitation. Here, in *Me Medicine vs. We Medicine*, I've expanded the focus beyond commodification of the body and into such new biotechnologies as pharmacogenetics and neurocognitive enhancement. But as well as being tied together by property concepts such as the commons, my earlier work and this book are linked through the constant surprises thrown up by the commercialization of biomedicine and the difficulty of regulating that unpredictability.

Because I've always been an avid proponent of mutuality and interrelationship as the most important issues in bioethics, rather than the dominant concept of autonomy, it's very gratifying for me to be able to acknowledge how much I've gained from the generosity of others in writing this book. Thanks to other members of the Worldwide Universities Network "Biocapital and Bioequity" group (particularly Nik Brown, Cathy Waldby, Danae McLeod, Andrew Webster, Sigrid Sterckx, and Julian Cockbain), I've been able to report on unexpected and often counterintuitive developments in Me Medicine—for example, the international trade in umbilical cord blood. Greg McClennan and Susan Jim of the Institute for Advanced Studies at the University of Bristol were instrumental in encouraging and hosting the Biocapital network. I'm very grateful to them.

I owe a great debt to Dr. Amar Jesani, editor of the *Indian Journal of Medical Ethics*, for alerting me to the human papillomavirus controversy in India and for so promptly and patiently replying to all my requests for further information. Richard Moxon, emeritus professor of pediatrics at the University of Oxford, was invaluable in commenting on the influenza and MMR sections of chapter 6. I've also had a great deal of clarification and assistance on the issue of cord clamping from two leading obstetricians who've campaigned consistently against early clamping, David Hutchon and Susan Bewley.

My old college classmate Leslie Pickering Francis and her coauthor Margaret Pabst Battin pulled out all the stops to make sure I had access to their magisterial work on vaccination and pandemics, *The Patient as Victim and Vector*. I owe a great deal to their insights and to all my discussions with Leslie during her sabbatical in Oxford this past spring, right up to the very moment when she boarded the airport bus.

In 2011, I was very lucky to have been invited to the Tarrytown Meetings, described on their website as "an annual invitational convening of people working to ensure that new human biotechnologies and related emerging technologies support rather than undermine social justice, human rights, ecological integrity, democratic governance, and the common good." My deepest thanks go to Marcy Darnovsky of the Center for Genetics and Society in Oakland for this invitation, and to all her colleagues, including but not limited to Richard Hayes, Charles Garzon, Osagie Obasogie, Emily Smith Beitiks, and Pete Shanks, for all they did to make this stimulating gathering possible. At Tarrytown, I was privileged to have long discussions with many other people who've contributed a great deal of help to me in writing this book, including Dorothy Roberts, Jonathan Kahn, Michele Goodwin, Jeremy Gruber, Helen Wallace, Tina Stevens, Judy Norsigian, Sally Whelan, Marcia Darling, and David Winickoff. Sarah Sexton of the Corner House couldn't make the Tarrytown meeting the year I went but has helped me enormously over the years, including by putting me in touch with Korean WomenLink and telling me about the work of the Indian Sama organization.

Through my association with the HeLEX Centre for Health, Law, and Emerging Technologies, at the University of Oxford, I've been assisted in all kinds of practical and intellectual ways by my colleagues Jane Kaye, Imogen Holbrook, Charles Foster, Jonathan Herring, and Nadja Kanellopolou. Imogen Goold of the Oxford Uehiro Centre for Practical Ethics has also been a great source of insights and information about the enhancement debate, as has my old friend Ruud ter Meulen of the Centre for Ethics in Medicine at the University of Bristol. I've also learned a great deal in discussions with Roger Brownsword, Stuart Hogarth, Heather Widdows, Gisli Palsson, Mary Fainsod Katzenstein, Peter Katzenstein, Alan Ryan, Andrea Boggio, Helen Busby, Mike Parker, and Richard Huxtable.

No other academic contributed more to the early stages of this book than Carol Sanger, who was a constant source of concrete support. I'm also grateful to Jennifer Merchant of the University of Paris–2, another "early adopter," for her enthusiasm and practical assistance, and to another very old friend, Castle Freeman Jr., for his suggestions and help.

My agent in New York, Michelle Tessler, contributed not only the book's subtitle but also a great deal of effort and goodwill. I want to thank her too, along with Patrick Fitzgerald, my editor at Columbia University Press, and Bridget Flannery-McCoy, my assistant editor. Except for one thundery late July day together in the city, Patrick, Bridget, and I have had to work at a transatlantic distance, but if anything it seems to have deepened our collaboration.

I'd like to dedicate this book to the memory of my grandmother, Lillian Esp Dickenson (1905–1971), and that of my father, Donald Moody Dickenson Jr. (1925–1951). As Quebec's motto says, *je me souviens*.

Donna Dickenson
Oxford, July 2012

ME MEDICINE vs. WE MEDICINE

1

A REALITY CHECK FOR
PERSONALIZED MEDICINE

We are in a new era of the life sciences, but in no area of research is the promise greater than in personalized medicine.
—Barack Obama, as a senator introducing the bill that became the Genomics and Personalized Medicine Act 2007

THE SOARING PROMISES MADE BY ADVOCATES OF PERSONALIZED medicine are probably loftier than those in any other medical or scientific realm today. In addition, the range of therapies covered by personalized medicine is even greater than then-Senator Obama realized. Direct-to-consumer genetic testing, personal tailored drug regimes, private umbilical cord blood banking, and "enhancement" technologies all come under that rubric. Part of this book's own promise is to introduce you to personalized medicine's lesser-known variants, illustrating how they all chime together in their hymns and psalms in praise of what I call "Me Medicine."

Sometimes, the clarion calls for these new technologies are delivered with almost messianic fervor, as in the case of this paean from Francis Collins, a former codirector of the Human Genome Project:

We are on the leading edge of a true revolution in medicine, one that promises to transform the traditional "one size fits all" approach into a much more powerful strategy that considers each individual as unique and as having special characteristics that should guide an approach to staying healthy. Although the scientific details to back up these broad claims are still evolving, the outline of a dramatic paradigm shift is coming into focus. . . . You have to be ready to embrace this new world.[1]

Do I? Why? I'd like to see more evidence before I decide. It's not that I'm afraid of new biotechnologies—I've spent my working life analyzing them and their ethical implications. Nor is it because I don't necessarily believe the promises will come true, although there are good reasons to doubt that they will ever really amount to a "dramatic paradigm shift."

Certainly, vast sums are pouring into personalized medicine: plans to spend $416 million on a four-year plan were announced in December 2011 by the National Institutes of Health,[2] and interest from the private sector is also intense. But the Human Genome Project (HGP) was also very generously funded, without having so far produced correspondingly weighty results for translational medicine, even a decade after it was announced that the human genome had been fully sequenced.[3] "Indeed, after 10 years of effort, geneticists are almost back to square one in knowing where to look for the roots of common disease."[4] Productivity in drug development actually *declined* after the HGP announced its completion, as did new license applications to the Food and Drug Administration.[5]

And we've been here before: other supposed "paradigm shifts," including gene therapy and embryonic stem cell research, haven't yet translated into routine clinical care either. Likewise for personalized medicine, current genetic tests and molecular diagnostics only apply to about 2 percent of the population, according to a March 2012 report from United Health's Center for Health Reform and Modernization.[6] A Harris poll of 2,760 patients and physicians in January and February 2012 indicated that doctors had recommended personal genetic tests for only 4 percent of their patients. This is hardly the stuff of a paradigm shift, at least not yet.[7] Some experts call the genomic revolution merely a "myth," arguing that at most we're witnessing a process of incremental change, one consistent with past trends in diagnostic innovation.[8]

Yet despite the lack of substantial evidence that personalized genetic testing is actually having a huge effect, the publicity around it may well be doing so—not necessarily for the best. I'm concerned that *Me Medicine* is eclipsing what I call *We Medicine*, so that we're losing sight of the notion that biotechnology can and should serve the common good. In my view, we would be wrong to prioritize personalized health technologies at the expense of public health measures, which have brought us comparative freedom from the ill health that plagued our ancestors. I see a pattern here—not only a similarity among all the apparently disparate forms of personalized medicine but also a familiar political formula: "private good, public bad."

Personalized medicine consciously appeals to the idea of the individual making free choices about her health, but in a much more sophisticated way than the simplistic stereotypes about free markets in healthcare versus welfare states, which were played out to tiresome length in the debates over the Patient Protection and Affordable Care Act 2010. Because it's much more palatable medicine—excuse the pun—it may not look like it's even part of that debate at all, but it is. If we take the Me Medicine fork in the healthcare road, we can't simultaneously go down the We Medicine route—the road less traveled by, in Robert Frost's phrase.

For example, there's been considerable growth in private umbilical cord blood banks, which charge a fee to store cord blood in an individual "account" for the newborn in the hope that stem cell technology will eventually allow the blood to be used as a sort of personal spare-parts kit. With one or two exceptions, these banks reserve the blood for the child's private use (Me Medicine), but there are also public cord blood banks (We Medicine) that actually achieve better clinical results.[9] Yet if enough parents bank their babies' umbilical cord blood privately, there won't be a sufficient supply for public cord blood banks, although those can be seen as both medically and ethically superior.

At the moment, perhaps surprisingly, the United States leads the world in the overall number of public cord blood banks. Despite our famous cult of individualism, we're tops in We Medicine there, but we won't stay that way if current trends toward private banking continue. Here and elsewhere, what may look like innocent individual consumer choices will shape how we as a society assure our health and that of future generations. So we need to think long and hard about how we want to prioritize the claims of Me and We rather than just hopping aboard the personalized medicine bandwagon like the great majority of commentators. This book is intended to let you make up your own mind about how you see those priorities, by giving you accurate, up-to-date medical and scientific evidence and locating the new technologies in their ethical and political context.

First, however: what exactly are these new personalized technologies, and how can they make such grand claims? Unlike this book, most works treat the various aspects of personalized medicine as separate developments, with different diagnoses and prognoses. The various techniques do at first look disparate. *Direct-to-consumer genetic testing*, in which a limited selection of genetic analyses are performed on a sample of saliva or a cheek swab, is probably the most familiar of the Me Medicine technologies. The field, which includes a number of "big players," such as the California company 23andMe, has been widely publicized by journalists who tried "retail genetics" out for themselves.[10] Along with the example of *private cord blood banking*, another increasingly familiar example of Me Medicine is *pharmacogenetics* or *pharmacogenomics*. Here, genetic typing is used to determine a patient's probable response to drugs, such as cancer treatments, and to tailor the pharmaceutical regime personally. Although, as we've seen, the percentage of patients undergoing such genetic diagnostics and treatments is still in the low single figures, *chemical* or *neurocognitive enhancement technologies* are even further away from everyday clinical practice, although they too have provoked column inches about what one of their most prominent opponents calls "the case against perfection."[11]

What do all these apparently disparate technologies have in common? Essentially, they're linked by two largely unchallenged assumptions: that "individual" is better than "social" and that we're on the cusp of a "true revolution in medicine" to make it more individualized. But are these assumptions justified? They may or may not be—that's what we'll discover as we go along—but the really interesting question is why so few have challenged them. The bookstores are full of somewhat dewy-eyed and often uncritically "pro" books about personalized medicine, such as Misha Angrist's *Here Is a Human Being: At the Dawn of Personal Genomics*; Francis Collins's *The Language of Life: DNA and the Revolution in Personalized Medicine*; Kevin Davies's *The Thousand-Dollar Genome: The Revolution in DNA Sequencing and the New Era of Personalized Medicine*; Thomas Goetz's *The Decision Tree: Taking Control of Your Health in the Era of Personalized Medicine*; Eric Topol's *The Creative Destruction of Medicine: How the Digital Revolution Will Create Better Health Care*; and Lone Frank's *My Beautiful Genome: Exploring Our Genetic Future, One Quirk at a Time*. But the book you're reading now doesn't take a knee-jerk "anti" position; it just aims to be balanced.

We need to ask why so many multinational firms, researchers, and—yes—presidents of the United States have all bought into personalized medicine. We urgently need a disinterested and balanced critique of personalized medicine's origins, the commercial interests that lie behind it, and the dynamics of its marketing as what I term *retail therapy*, that is, medical treatment and diagnostic regimes conceived as consumer goods. Just as the body itself has been commodified—the argument of my previous book, *Body Shopping*—so medicine is increasingly seen as a commodity, in both insurance-based and more socialized healthcare systems.

Historically, it was not Me Medicine but We Medicine—programs like public vaccination, clean water, and screening for tuberculosis—that brought us reduced infant mortality, comparative freedom from contagious disease, and an enhanced lifespan. Yet today, many of these public programs seem to be increasingly distrusted, even detested. Some U.S. campaigners against the measles, mumps, and rubella (MMR) vaccine have allegedly accused physicians who administer the vaccine of being in the same league as Nazi concentration camp doctors.[12] Vaccination programs are in profound trouble in many parts of the world. In India, a similar though less virulent reaction has arisen against what might seem like a model public health campaign, the vaccination of young girls against the human papillomavirus implicated in cervical cancer.[13] In Muslim areas of northern Nigeria, a country which accounts for about 45 percent of polio cases worldwide, a World Health Organization vaccination campaign was boycotted as a Western plot to spread HIV and AIDS through adulterated injections.[14]

In contrast, when a new medical development combines scientific mystique and the wand-waving word "personal," the reaction worldwide will probably be overwhelming adulation. That was very much the case when the Korean researcher Hwang Woo Suk announced in 2005 that he had successfully created eleven "patient-specific" stem cell lines. Hwang was pointing toward the possibility that eventually everyone could have a personal spare-parts kit, overcoming the problem of immune rejection when organs are transplanted.[15] "After Hwang's article was published, he turned into a sacred figure."[16] The reaction, in both East and West, was so euphoric that Hwang offered to set up a worldwide franchise of his method, with satellite laboratories in California and England—before his claim was revealed to be totally false. He hadn't created a single successful cell line, even though he had published his "findings" in the prestigious journal *Science*—fooling both the editors and the scientific world at large.

But how was that possible? Although it's a bit speculative, perhaps one reason is the spell cast by the idea of personalized therapy. Some of that unconfined joy and uncritical adulation had a genuinely scientific appeal—that is, if the technique had worked, and if it hadn't required dangerous levels of hormonal stimulation to produce the human eggs that the technique demanded in huge quantities.[17] But it also seems plausible that Hwang's supposedly patient-specific stem cells appealed because they pushed the right buttons in our psyches: the ones marked "personal" and "individual." The possibility of a commercial franchise mooted by Hwang before his unmasking indicates that pushing those buttons is also important and attractive to corporate interests.

It's hard to explain why else ferreting out the truth took determined campaigning by a not-very-well-known Korean feminist group, Korean Womenlink, and the subsequent acknowledgment by Hwang's principal colleague, Gerald Schatten, that the methods used in sourcing the eggs had been ethically dubious, eventually leading to a recognition of the scientific inaccuracy of the claim. It's also difficult to understand why more attention wasn't paid to improving the rate of tissue rejection through further advances in the already promising field of immunology, as a few scientists did argue at the time.[18] That would mean that we could recruit a wider range of tissue donors without having to worry about tissue matching, to avoid rejection of the transplant, or the alternative of heavy and risky doses of immunosuppressants. We could concentrate on practical methods of improving the success of altruistic donation from others rather than on our own speculative personal spare-parts kits.

But that's the dull alternative of We Medicine, isn't it? How can it compare with the exciting promise of personalized medicine? Here's the story of someone who did test that promise at no little risk to himself. Like Collins, he's one of the new "evangelists" of Me Medicine. His story might help us begin to piece together the reasons why so many observers have joined that new movement. There's also a dominant theme of threat running through his story, which I will consider later in this chapter and explore throughout this book as one possible hypothesis explaining the rise of Me Medicine.

THE NEW EVANGELISTS

In writing his 2009 book *Experimental Man*, David Ewing Duncan—the chief correspondent of National Public Radio's "Biotech Nation" and director of the Center for Life Science Policy at the University of California, Berkeley—had himself tested for 320 chemical toxins and up to ten million genetic markers. He spent twenty-two hours having magnetic resonance imaging and underwent the drawing of 1.7 liters of blood. The total cost of all the tests that Duncan endured was between $150,000 and $500,000. That's the range Duncan himself gives, which seems more than a little vague, but many of the tests were supplied gratis by the genetic testing industry. Whichever end of the dollar scale turns out to be most accurate, he still consumed a great deal of medical resources.

Although you might think that there's nothing particularly liberating about being an experimental guinea pig on such a scale, Duncan urges readers of the book and visitors to his website to sign what he calls a "Personalized Health Manifesto": "an old-fashioned call to arms and action plan for a new age of health care."[19] We heard the same campground-meeting rhetoric from Francis Collins. Personalized medicine seems to be becoming the equivalent of nineteenth-century American revivalism.

Back in the 1840s, when the students at Mount Holyoke seminary were called on by their college president Mary Lyon to stand up and testify to their desire to lead a Christian life, the young Emily Dickinson was one of the few who remained in her seat. "They thought it queer I didn't rise," she remarked afterward. "I thought a lie would be queerer." Similarly, Duncan reportedly called on attendees at the U.S. National Undergraduate Bioethics Conference in 2011 to demonstrate their conversion to personalized medicine with a show of hands. Only one modern-day Dickinson's hand remained down. "Too bad," Duncan reportedly said. "It's happening anyway."[20] To be fair, Duncan actually concludes in his book that the direct-to-consumer genetic tests he tried are mostly disappointing. He advises not placing too much reliance on the results—yet. But when the science is perfected, his reasoning seems to run, what's not to like?

To start with, that "when" has every appearance of being an "if," although many proponents of personalized medicine make very big claims indeed. It's been asserted that a baby could have her genome fully sequenced at birth, along with her susceptibility to particular diseases. She could then enjoy the benefits of made-to-order diagnostic tools and drugs throughout her lifetime.[21] That really is the Holy Grail of personalized medicine, but it makes huge and currently unfounded assumptions about how much genetic and genomic medicine is actually able to predict. Most major diseases are caused by the interplay of many genes rather than one, and they arise from both environmental and genetic causes.[22]

Proponents of personalized medicine's benefits point with some justification, however, to the evolving area of biomedicine known as pharmacogenetics or pharmacogenomics. For example, the drug warfarin is an oral anticoagulant commonly used to prevent or manage venous thrombosis (clotting). It's sometimes difficult to determine the correct dosage for an individual patient: thinning the blood excessively can be an unwanted side effect, carrying its own risks. But now, warfarin dosage can be tailored to identify particular patients at increased risk of bleeding by sequencing two genes that account for most of the variation in how people react to the drug. Even here, there's some skepticism about whether pharmacogenetics has actually improved outcomes for patients[23] and whether more extensive reliance on personally tailored drug regimes requires a "leap of faith,"[24] as I'll discuss at greater length in chapter 3. Nonetheless, if sufficient evidence were amassed to show that pharmacogenetic dosage of warfarin is clinically effective, this would exemplify one meaning of personalized medicine that does seem genuinely beneficial: drug treatment tailored to the patient on an evidence-based model for better clinical care.

Whether that's really personalized in the sense of *individualized*, however, is arguable. In the warfarin example, individuals are classified into *groups* according to which allele (variant) of the relevant gene they have. It might be better called "small-group medicine," though that's nowhere near as catchy. Personalized medicine in the warfarin example is still more "We" than "Me," even though warfarin is frequently cited by Me Medicine advocates as proof that truly individualized medicine is already a reality.

Even the biotechnology industry–linked Personalized Medicine Coalition concedes that pharmacogenetics is about population subgroup response to particular drugs. It's still an improvement, they argue, because only 50 percent of the population responds to a typical drug—a figure that can be translated into a higher probability through pharmacogenetics.[25] Of course, this is an improvement, but it's still not really *personalized* medicine: a probability applies by definition to a statistical *group*. The phenomenon of statistical independence means that no probability can tell you with certainty that you as an *individual* will or will not respond to a drug, any more than the 50 percent probability that a coin toss will come up heads can predict whether the next toss will come up tails.

Evangelists for personalized medicine often adduce the discovery of blood types as the pioneering example of individualized care. Yet that, too, is about assigning individual patients to serum groups (A, O, B, and AB), which are further divided into subgroups by rhesus type (positive or negative). The discovery of blood groups did revolutionize transplant surgery, and so it could count as a genuine example of a paradigm shift—but whether it's truly "personalized" is arguable.

Advocates of personalized medicine frequently play on the stereotype that traditional medicine ignores our individuality. For Eric Topol, hidebound conventional therapies require "creative destruction" in favor of a genuinely individualized medicine:

> This is a new era of medicine, in which each person can be near [*sic*] fully defined at the individual level, instead of how we practice medicine at a population level, with mass screening policies for such conditions as breast or prostate cancer and use of the same medication and dosage for a diagnosis rather than a patient. We are each unique human beings, but up until now there was no way to establish one's biologic or physiologic individuality.[26]

Likewise, the Personalized Medicine Coalition asserts that "physicians can now go beyond the 'one size fits all' model of medicine to make the most effective clinical decisions for individual patients."[27] Francis Collins used similar language when he predicted that personalized medicine will "transform the traditional 'one size fits all' approach into a much more powerful strategy that considers each individual as unique."[28]

Yet good practitioners have always relied on close observation of the particular patient. As Hippocrates said, "It is far more important to know what person the disease has than to know what disease the person has." The notion of "whole-person treatment" didn't originate with pharmacogenetics or direct-to-consumer genetic testing. Indeed, as Collins himself admits, taking a family history, that staple of old-fashioned medical practice, still reveals risk proclivity for particular diseases more accurately than consumer genetics.[29] And looking at the family, by definition, means moving beyond the individual, from Me to We.

So it seems fair to say that personalized medicine is nowhere near as new or innovative as it claims to be—nor as successful. Direct-to-consumer genetic testing, for example, is likely to yield conflicting results because the methods are not standardized and the disease probabilities are not universally accepted by experts. These "retail genetics" firms test for forms of genetic information (single nucleotide polymorphisms, or SNPs, single-letter differences in DNA between individuals), but none of them tests for the same set of SNPs. Much to his consternation, David Ewing Duncan received three frantically different assessments of his heart attack risk from three different genetic testing companies. The director of deCODEme, Kari Stephansson, even telephoned him personally from Iceland to urge him to start taking cholesterol-lowering statins right away—but the other tests had rated him at medium or low risk of developing dangerously high cholesterol. As Duncan puts it in a laconic chapter subheading, "I'm doomed. Or not."

Yet Duncan remains an ardent advocate of personalized medicine. Even more critical observers tend not to go beyond the biomedical reasons for doubting whether personalized medicine really has a future.[30] I'm not dismissing those medical and scientific doubts: they are valid and valuable. The recommendations of medical professional bodies, like the evidence-based judgment on DTC genetic testing of the American Society for Clinical Oncology,[31] are entirely appropriate to the task and competence of the observers. But for this book's purposes—a comprehensive and skeptical survey of all the various trends toward Me Medicine—we need to go further.

Let's break out of the biomedical box and introduce four wider social, political, and ethical reasons why people might be tempted to buy into personalized medicine: (1) threat and contamination, (2) narcissism and the "bowling alone" phenomenon, (3) corporate interests and neoliberalism, and, finally, (4) choice and autonomy. After a preliminary appraisal here, these four possible hypotheses will be evaluated against each of the specific medical developments examined in successive chapters. By the end of the book, we should have a much clearer idea of the profound social and political reasons why Me Medicine threatens to edge out We Medicine and a rational program for doing something about it, if that's what we decide is appropriate.

Four Approaches to Understanding the "Me Medicine" Versus "We Medicine" Phenomenon

1. Threat and contamination
2. Narcissism and "bowling alone"
3. Corporate interests and political neoliberalism
4. The sacredness of personal choice

THREAT AND CONTAMINATION

In his book *Experimental Man*, subtitled *What One Man's Body Reveals About His Future, Your Health, and Our Toxic World*, David Ewing Duncan reveals that his testing program—testing in all senses—was motivated not just by intellectual curiosity but by a sense of threat and contamination. Unbeknownst to his mother, an environmental activist, Duncan spent his idyllic Kansas boyhood wading in streams full of chemical runoff or mining the "mother-lode" of a landfill site for old bottles, broken machines, steering wheels, and, as it turned out, heavy metals. Brought up to believe that he came from a family of long-lived individuals, he describes feeling fragile for the first time when he discovers that his genes can't protect him against the abnormally high levels of toxic residues in his blood. In a circular and ironic relationship with threat, the Experimental Man project that he underwent to take control of his health actually left him feeling more at risk than ever before.

That's one sense of threat, but there are also others that might help to explain the rise of personalized medicine. Contamination and pollution as powerful motivating fears can, of course, extend to many forms of "dirt" and impurity.[32] The UK system of altruistic blood donation is increasingly being bypassed by people wanting to bank their own blood for future use. Frightened by possible contamination of communal blood banks by HIV and BSE (bovine spongiform encephalopathy, or "mad cow disease"), patients scheduled for operations may now choose to avoid that threat by banking their own blood in advance.[33] Once the epitome of We Medicine, that marvel of efficiency and altruism depicted by Richard Titmuss in his influential book *The Gift Relationship*, the UK national blood service now risks being transformed into a form of Me Medicine. The model for blood use would then become one of depositing in a personal account rather than donating to or drawing on a communal resource.

Personal or "autologous" blood depositing is still only practiced in a minority of cases: the patient must be healthy enough to withstand not only the procedure but also the withdrawal of blood beforehand. But more people would do it if they could. A Eurobarometer survey of European public opinion found that 25 percent of respondents would only accept their own blood if they needed a transfusion. Another 23 percent would also be willing to take blood from a known person such as a friend or relative, though not from a stranger.[34] That brings the total who want nothing to do with communal blood up to roughly half the European survey population: powerful evidence of a growing sense of threat and contamination in what was once seen as the quintessential symbol of social solidarity, blood donation.[35]

In the United Kingdom, the Factor VIII hemophiliac controversy and the emergence of an untreatable variant of Creutzfeldt-Jakob Disease (CJD, a form of dementia possibly linked to mad cow disease) do at least give patients some reason to fear a threat from communal blood. But we'll see in chapter 4 that private umbilical cord blood banking for an infant's personal future use is also on the rise, although the actual evidence indicates that rather than reducing the threat of danger, it may actually pose a risk to the baby.[36] Yet *perceived* threat is highly relevant, as is evident in the lengthy but sometimes inaccurate lists of diseases from which private banks claim the baby can be protected by banking the blood.

The link between toxin threat and personal genetics was consciously built into the Human Genome Project itself, the British geneticist Helen Wallace maintains.[37] Using documents obtained through litigation, she's produced extensive evidence, which I'll examine critically in the next chapter, purportedly demonstrating that Big Tobacco threw itself into funding genetic and genomic research in the hope of narrowing down those who were "genetically susceptible" to tobacco smoke, thus reassuring the majority of the population that they were at no risk from smoking.

Wallace claims that the tobacco industry even promoted the idea that an unknown gene both drove particular people to smoke and made them genetically vulnerable to carcinogens in cigarettes. No such genetic basis for wanting to smoke or for being particularly susceptible to smoking ever materialized, of course. But the notion of splitting off certain vulnerable individuals, the framing of smoking as a consumer choice, and the background sense of threat all fit uncomfortably neatly into the pattern of Me Medicine.

These examples all draw on physiological threats, but it might well be said that the current state of healthcare leaves us all feeling threatened for financial or political reasons, such as spiraling costs, the difficulty in finding insurance, and the reluctance of many family doctors to take on new Medicare patients. Even in the United Kingdom and elsewhere in Europe, austerity cuts mean that unified and universal healthcare is increasingly under threat. Although the United Kingdom still formally retains the National Health Service, in March 2012 a government-sponsored bill, condemned by medical professional bodies, introduced radical new provisions that have been criticized as likely to lead to "cherry-picking" of better-off patients and neglect of the less wealthy.[38] In April 2013, responsibility for public health—We Medicine—is to be transferred from the unified National Health Service to cash-strapped local authorities, who may not all be able to provide the same level of service.[39] So here, too, threat is a dominant motif, possibly leading British patients to feel that in future they'll have to take charge of their own health to a greater degree, "topping up" their NHS coverage with personal insurance plans and establishing their individual genetic risks for certain diseases.

Yet it also seems possible that personalized medicine *itself* could produce new kinds of threats, and thus patients would simply be exchanging rather than eliminating forms of risk. For example, if patients are ranked pharmaco-genetically according to how well they're likely to respond to expensive drugs, those less likely to respond may well be denied treatments that they would have received on a "one-size-fits-all" model of prescribing.[40] This is the down-side of what is more commonly presented as a major advantage of pharmaco-genetics: that tailored drugs will "spare expense and side effects" for those who are genetically less likely to benefit from a particular treatment.[41]

Given that its most ardent defenders present containment of rising medical costs as a major attraction of personalized medicine, we can assume that this is indeed high on the agenda. Just as those who believe in reincarnation typi-cally think that in a previous life they were emperors rather than galley slaves, so might we all think we will be among the genetic elite who will get the en-hanced new products of pharmacogenetics. But what if we're among the new untouchables instead?

In all these circumstances, it's natural to feel that you're going to be on your own if you fall ill and that it makes sense to try to forecast and minimize your risk by finding out all you can about your genetic propensity to particular dis-eases. Do-it-yourself genetic testing, for example, is presented as one means to that end. Sometimes firms play up the risk-minimization angle quite directly: for example, a DTC firm offering to rate young adults' sports abilities by ge-netic proclivity has been accused of playing on scare stories about deaths in young athletes.[42] More frequently, however, DTC firms present themselves as "empowering" their customers, hijacking the rhetoric of the 1960s.[43] For ex-ample, 23andMe's website asserts: "The company was founded to empower in-dividuals and develop new ways of accelerating scientific research."

The virtuous twin of threat might appear to be promise, upon which Me technologies such as neurological or genetic enhancement clearly play. But it's worth noting that the promises made by enhancement are for individuals or a comparatively small elite: they will never be mass technologies. Indeed, that designer cachet might be part of the sales pitch. That brings us to a second possible explanation for the rise and rise of Me Medicine: narcissism.

NARCISSISM AND "BOWLING ALONE"

23andMe, Knome, deCODEme, and MyGenome: is it only a coincidence that the words "me" and "my" are part of the brand name for so many DTC genetic testing companies?[44] Or is retail genetics part of a more generalized trend toward narcissism and self-absorption? "No single event initiated the narcissism epidemic; instead, Americans' core cultural ideas slowly became more focused on self-admiration and self-expression. At the same time, Americans' faith in the power of collective action in the government was lost."[45] Jean Twenge and Keith Campbell, authors of *The Narcissism Epidemic*, remark on the use of "I," "me," and "my" as branding devices outside biomedicine—notably the repetition of "I" in the iPod, iPhone, and iPad. (Even if the "i" is in lowercase, it's still all about "me.") David Ewing Duncan actually suggests that eventually we will each own a handheld device, which he jokingly but appropriately terms an "iHealth." On it, he predicts, we'll track our genomes and most recent scans, inputting environmental data as we go through the low-tech drudgery of everyday life.[46]

The concept of a narcissism epidemic isn't strongly medical or scientific, although Twenge and Campbell do produce evidence of a recent rise in narcissistic personality traits on psychological profile tests taken by college students. Mainly, however, they delineate a sense of entitlement that has permeated popular culture and has changed child-rearing practices to overemphasize the child's intrinsic specialness, at the expense of an awareness of others' needs. Twenge and Campbell reserve particular scorn for notions about needing to love yourself first before you can love anybody else. As an NBC public service announcement puts it, "You may not realize it, but everyone is born with their one true love—themselves."[47] Narcissism in this sense is different from individualism—and more pernicious.

> America has always been an individualistic nation, but it was focused on ideas of individual liberty, freedom from tyranny, and fundamental equality—values that emphasized independence, not narcissism. But when these powerful ideas were supplemented by the new values of self-admiration and self-expression, the results were ugly.[48]

Although Twenge and Campbell don't make the connection to DTC genetic testing, they do argue that the Internet—on which retail genetics depends—promotes narcissistic behaviors, such as endlessly refining your MySpace page or fattening up your list of Facebook "friends" to emphasize quantity rather than quality of interactions. You could also see personalized medicine, particularly retail genetics, as a response to celebrity culture. Acres of genetic analysis all about your individual genome, the extra option of an ancestor-tracing service offered by some DTC firms, the chance to join a social network of other customers offered by some services—all these features could well make purchasers feel that they're as newsworthy as celebrities and that their body's idiosyncrasies are the stuff of drama. Narcissism might go hand in hand with the "genetic mystique" described by Dorothy Nelkin and Susan Lindee in *The DNA Mystique: The Gene as Cultural Icon*: "Just as the Christian soul has provided an archetypal concept through which to understand the person and the continuity of the self, so DNA appears in popular culture as a soul-like entity, a holy and immortal relic. . . . It is the essential entity—the location of the true self—in the narratives of biological determinism."[49] The genetic mystique is intertwined with the idea of "genetic exceptionalism," the implicit assumption that genetics and genomics reveal more profound truths than other sciences. Normally, these two concepts accompany genetic determinism—the proposition that genes determine our behavior, as found in media articles claiming that scientists have discovered genes determining everything from voting patterns[50] to becoming a ruthless dictator.[51]

In the case of retail genetics, however, the marketing is predicated not on genetic determinism but rather on its opposite: an underlying assumption that we are in control of our behavior so that we can alter unhealthy eating or exercise patterns, for example, to counter a genetic predisposition to heart disease. (We'll see in chapter 2 that these promises, however, are more honored in the breach than in the observance.) When combined with the "ideology of wellness,"[52] geneticization means that those who are well can then take credit not just for their superior genes but also for their initiative in counteracting any "inferior" ones.

It's odd to see genetic exceptionalism divorced from genetic determinism: the more usual assumption is that genes dictate not only who we are but also what we do. Yet genetic exceptionalism is actually strengthened by avoiding the incoherence of genetic determinism. After all, do your genes dictate that you believe that your genes dictate what you believe?

The notion that retail therapy plays on narcissism and the genetic mystique seems initially plausible, whether or not that sense of narcissism is growing as exponentially as Twenge and Campbell assert. It also accords with the analysis in Robert Putnam's influential book *Bowling Alone*. Putnam argues that in the last third of the twentieth century, political, civic, and religious participation all declined, along with voluntarism, trust, and reciprocity. A sense of "we"-ness went missing in just thirty years, he says, and was replaced instead by the virulent culture wars that still dominate American politics.

The dominant theme is simple: For the first two-thirds of the twentieth century a powerful tide bore Americans into ever-deeper engagement in the life of their communities, but a few decades ago—silently, without warning—that tide reversed and we were overtaken by a treacherous rip current. Without at first noticing, we have been pulled apart from one another and from our communities over the last third of a century.[53]

Putnam particularly contrasts the "civic-minded" World War II generation with their supposedly more self-centered children, the generation that came of age in the 1960s and 1970s.[54] This seems to me an overworked and inaccurate comparison. The so-called Greatest Generation may well have been heroic in wartime, but in many cases the trauma of their experiences left them cold and inward-turned. Michael Cunningham's novel *The Hours* captures the claustrophobia of the 1950s well, with its fraught description of a postwar housewife's suicide attempt as an escape from her stifling and conformist family life. Toni Morrison, whose novel *Home* likewise desentimentalizes the period, has described her urge to "take the scab off the 50s, the general idea of it as very comfortable, happy, nostalgic . . . Oh, please."[55]

Putnam admits that the 1950s weren't such a golden age for African or Hispanic Americans, but he insists that at least for whites, "engagement in community affairs and the sense of shared identity and reciprocity had never been higher."[56] Yet even if he's right about the quantity of civic engagement, that doesn't say anything about its quality: the uses served by that togetherness. Core civic organizations of the 1950s included the likes of the Masons and the Daughters of the American Revolution: one resolutely excluding women, the other recruiting women to protest against any form of progressive policy. At the extreme, Putnam acknowledges that Ku Klux Klansmen also engage in community affairs and share a sense of identity. As Grand Wizard Jeff Coleman said, "Really, we're just like the Lions or the Elks. We want to be involved in the community."[57]

By contrast, the stereotype of the 1960s generation as hedonistic hippies leaves out the Vietnam draft resistance, women's liberation, and the civil rights movement. Many, such as Medgar Evers, the three "Freedom Riders" killed in Mississippi in 1964, and, of course, Martin Luther King, died or went to prison for those collective causes. It demeans their memory to label the entire 1960s generation as self-centered individualists. And if that's true, then the notion of remorseless decay from a golden age of immediate postwar togetherness becomes less plausible.

The Narcissism Epidemic and *Bowling Alone* are both premised on the claim that "social capital"—a set of connections among individuals and norms of reciprocity—is in grave decline. Yet as we've seen, Putnam does recognize that social capital is good for the "in" group but may redound against outsiders. On the logical principle of the excluded middle—the proposition that one factor cannot explain both an effect and the absence of an effect—the social capital explanation runs into difficulties. Does it create a greater willingness to help the socially excluded, or can it actually result in group closure against the dispossessed?

This paradox has important implications for Me Medicine. On the one hand, the supposed *decline* in social capital could explain the current focus on "Me"-ness: the feeling that you're responsible for your own health. That would be the implication suggested by *The Narcissism Epidemic* and *Bowling Alone*. It might also be an explanation for the rocketing growth of cosmetic surgery. In the words of Martha Hennessey of the Catholic Worker movement, "Americans have retreated into collective narcissism."[58] The only thing we do collectively, in this view, is to agree that we're allowed to focus entirely on ourselves.

However, more "Me"-ness in medicine and social policy generally could also be explained by an *increase* in social capital, one developed through group closure. It all depends on what the group stands for, not on the mere fact of its being a group, and on how it attains its unity and purpose. Since writing *Bowling Alone*, which gave the impression that social solidarity was a terminal case, Putnam has praised the teams who canvassed for President Obama in the 2008 Democratic primaries as harbingers of a revived sense of bowling together. That's all well and good, but their nemesis, the Tea Party, is also a grassroots grouping, albeit one with significant support in high financial places.[59] The sense of shared identity among Tea Party members likewise depends on a common platform, which they would see as self-reliance and independence. It also depends on rallying the troops against the opposition, defined variously as immigrants, Washington bureaucrats, bankers, or political leftists, in a collective fashion, although in the paradoxical name of individualism.

It's been suggested that Americans are most likely to vote in favor of redistributive social and health programs if they see "people like us" as the beneficiaries.[60] This nasty side effect of group identification isn't softened by community integration; rather, the reverse is true. "The greater the racial and ethnic diversity of the community, and the more likely it is that voters see their tax dollars going to assist 'the other,' the lower the support for any spending, be it on health, schools or welfare."[61] When a federal healthcare program with overtones of We Medicine is proposed by a president who embodies "the other" to some white Americans, you might speculate that a wholesale flight into Me Medicine is only to be expected.

While psychological factors like group identity or narcissism are intuitively plausible explanations of the rise of Me Medicine, they only take us so far. In fact, they actually produce contradictory predictions about whether We or Me Medicine is likely to result from a decline in communal identity. Let's consider another possible hypothesis: corporate interests and political neoliberalism.

CORPORATE INTERESTS AND POLITICAL NEOLIBERALISM

Should scientists see themselves as part of a worldwide NGO [nongovernmental organization], upholding a set of shared values? I think that's exactly the way they used to be in previous centuries . . . and actually that international fellowship is by no means gone, but it's threatened when people try to walk both sides of the line, mingling scientific contribution with profit-making activity. . . . We in Western society are going through a period of intensifying belief in private ownership, to the detriment of the public good. Individual selfishness is held up as the best way to advance civilization, and through the process of globalization these beliefs are being exported to the world as a whole, making it not only less just but also less safe.[62]

In this quotation, the Nobel prize–winning geneticist John Sulston sounds at first as if he's saying that scientists are becoming solitary bowlers or selfish narcissists. Actually, he's criticizing the way in which the "business model" of science is changing from public to private benefit: what he calls "an intensifying belief in private ownership, to the detriment of the public good." Sulston doesn't just present this transformation in atomized individual terms, nor does he see it as primarily psychological, in the way of the narcissism model. Instead, he's suggesting that "through the process of globalization," a political and economic transformation, science is moving from We to Me.

Personalized medicine hasn't just sprung up in a political or economic vacuum. It has coincided with the ascendancy of "neoliberal" political ideology, which, as Sulston argues, has affected science and medicine profoundly.

This viewpoint isn't unique to Sulston: it is taken up and analyzed at considerable depth in Philip Mirowski's cleverly titled *Science-Mart: Privatizing American Science*. As a professor of both economics and philosophy of science, Mirowski is well qualified to track what he believes to be a deliberate political effort over the past four decades to incorporate neoliberal economic and political policies into academic science.

Neoliberalism

The package of economic and political measures known as neoliberalism typically includes the following policies:

- "Rolling back the state" through abolishing regulatory legislation and making stringent cuts in public spending, while simultaneously
- increasing the involvement of private corporations in key governmental functions, effectively privatizing areas of public provision such as education, health, and scientific research, thus
- transferring public wealth to private corporations through awarding monopoly contracts and outsourcing necessary services. The underpinning rationale is
- viewing markets as the only necessary form of discipline in any economy, in the belief that markets automatically correct their own mistakes, while simultaneously using public-sector funds to subsidize loss-making private activities. All these policies are premised on
- downplaying the notion of the public good or even denying that there is any such thing.

Neoliberalism—also known variously as "free-market economics," "globalization" (the term used by Sulston), or the "Chicago school," after the university associated with its leading exponent, Milton Friedman—gained political ascendancy in the United States and United Kingdom during the early 1980s. It is distinguished from nineteenth-century liberalism by its politically conservative tendencies: John Stuart Mill's liberalism was in some ways quite radical, for example, in his proposals that women should gain the vote and enter Parliament. Modern neoliberalism, however, is associated with the "neoconservative" movement—although, confusingly, in American politics a "liberal" is someone of the moderate political left.[63]

The hallmark of neoliberalism is the belief that state intervention, and in particular the welfare state, is harmful to free markets, which are the true creators of wealth.[64] The influential political theorist Michael Sandel believes that "market triumphalism" is now so entrenched that rather than *having* a market economy, we quite simply *are* a market economy: markets control and define our society, with nonmarket moral values increasingly edged out.[65] Those values might well include those that Putnam praises: civic feeling, compassion, solidarity, altruism, and a sense that there is such a thing as the common good.

This dominance of the market is the source of the ideology of "private good, public bad," which I linked earlier in this chapter to the rise of Me Medicine and the decline of We Medicine. If the notion of common welfare is to be distrusted, and if interventions such as public health programs are regarded as interference with individual rights, We Medicine will automatically be suspect. Hostile reactions to vaccination programs, for example, aren't just a matter of a few vituperative cranks: they're sanctioned in an indirect way by a more general climate of distrust for any state initiative.

But although the official message of neoliberalism is "hands off," the actual policies pursued everywhere from banking to biotechnology involve state intervention to subsidize loss-making activity for the private sector. For banks, that's meant the losses made on junk bonds and subprime mortgages; for science, it's the non-profit-making research and development phases. In both cases, we often witness the conversion of the asset to private hands once it's profitable: what the sociologist Stuart Hall calls "siphoning state funding to the private sector."[66] In the UK banking sector, for example, the government rescued the failed bank Northern Rock with taxpayers' money, to avoid another collapse like that of Lehman Brothers in the United States. But it then overrode calls to keep the bank in national hands and sold it in November 2011 to Virgin Money, reportedly for something like half what it had paid for it.

In the United States, the Bayh-Dole Act of 1980 encouraged private capital to enter the scientific marketplace and promised to subsidize any losses incurred in the process. "To allow wealth from discoveries to be realized, the Act turned the principle of capitalism on its head: 'private risk yields private loss or gain' became 'public risk yields public loss or private gain'—a form of 'heads I win, tails you lose.'"[67] In April 2012, the Obama White House announced its "National Bioeconomy Blueprint," which "outlines steps that agencies can take to drive the bioeconomy" in a time of economic uncertainty, much in the spirit of Bayh-Dole.[68] Mention of any risks from genetic engineering or other technologies is confined to a footnote, otherwise framed as "beyond the scope of this document."[69]

We can trace this same neoliberal trajectory in the development of firms such as deCODE Genetics, which depended on the free public resource of the Icelandic national population database but retained all profits for itself.[70] It's also evident in the way that private umbilical cord blood banks in the United Kingdom often piggyback on NHS hospital staff provision and rely for their marketing appeal on the hope that stem cell research—typically funded by government research councils and thus by the taxpayer—will "add value" to the stored blood.

So it's not just a coincidence that personalized medicine has flourished at the same time that the majority of governments throughout both the developed and developing world—including India and China—are pursuing neoliberal policies. In many cases, the profitability of Me Medicine depends directly on those policies. At the highest governmental levels, public backing has been solicited to underpin private-sector profit making from biotechnology.

In Executive Order 13326 of September 2001, President George W. Bush established the Presidential Council of Advisors on Science and Technology (PCAST). This was a private-sector body with cabinet-level status—as if it were an arm of elected government. Its mandate was to "assist the National Science and Technology Council [the public body] in securing private sector involvement in its activities." Under President Obama, a new Executive Order, number 13539, reestablished PCAST on a less obviously proindustry footing but retained private-sector involvement. Its mission is now to "solicit information and ideas from the broad range of stakeholders, including but not limited to the research community, the private sector, universities, national laboratories, State and local governments, foundations, and nonprofit organizations."[71]

Where does the biotechnology industry see profits in personalized medicine? It's crucial to bear in mind the adage about capitalism not serving existing markets so much as creating demand where none existed before. Even the solidly middle-of-the-road Nuffield Council on Bioethics in the United Kingdom remarks of personalized medicine that "personalisation is sometimes represented as a response to demand, but in some cases at least it seems to be a case of supply looking for demand."[72] Private cord blood banking and retail genetics are both perfect examples of creating demand where none existed before. Who would have predicted twenty years ago that you could get people to pay to bank their infant's umbilical cord blood or to have a spit sample analyzed to predict their personal propensity to common diseases?

Even pharmacogenetics, which goes back further than either of those technologies and has a stronger evidence base, also demonstrates political and

economic elements. The PCAST report of 2008 states quite openly that industry's interest in pharmacogenetics is dictated not only by scientific developments but also by cost and market considerations. Because trial and control
groups can be genetically matched more closely, pharmacogenetics potentially reduces the size, cost, and duration of expensive clinical trials. With
over 70 percent of drug trials now performed in the private sector,[73] the drug
industry sees cutting trial costs as crucial to profitability (even though prominent critics such as Marcia Angell, the former editor of the *New England Journal of Medicine*, think that the industry actually spends far more on lobbying
than on product development).[74] The PCAST report also holds out the hope
for industry that failed trials might still work on subsectors of the clinical
population, so that research investment made in dud drugs wouldn't go to
waste.

Even more important, for a pharmaceutical industry facing the expiration
of patent protection on many of its best-selling drugs is finding new markets.
By breaking an existing medication down into different "size ranges" and persuading customers that they can't simply rely on a one-size-fits-all product,
pharmaceutical companies can create new niche markets. Similarly, private
umbilical cord blood banking taps into a huge potential market—all expectant mothers who can afford it, plus grandparents—with a "product" that no
one could have dreamed of before but that diligent parents and grandparents
may now wrongly believe is essential to the child's future health.

It would be even more advantageous for the pharmaceutical industry if the
individual patient could be persuaded to pay for genetic typing out of her own
pocket, so that she would then know which of the niche pharmaceuticals is
her "size." Although they're too imprecise at the moment to allow for that, and
while they test for only a fraction of relevant SNPs, retail genetic tests accus-
tom healthcare consumers to the idea of personalized drug regimes. In some
cases, the link isn't just psychological: it's much more direct. Before it shut
down its DTC service after being acquired in summer 2012 by Life Technologies, Navigenics had begun offering an additional pharmacogenetics service
for its existing DTC customers: "Genetic insights from Navigenics can help
you and your doctor select medications that may be right for your genetic
makeup."[75]

Now that the thousand-dollar whole-genome scan has become a reality,
customers could conceivably have all their personalized genetic information ready for access when needed, so that prescribing on a pharmacogenetic model could become much more commonplace. And if customers pick

up the tab for genetic testing, none of this will have cost the drug companies a bean. So the costs of diagnostics are beginning to be transferred from the public health system or insurers to the private individual while profits are transferred from the private individual to private companies. This process is quite consistent with the flow of income predicted by the neoliberal economic model.

Another potential source of profit from personalized genetic testing lies in biobanks—tissue and data repositories that can be sold to other firms or mined for research. Why are major retail genetic companies willing to sell direct-to-consumer tests at fire-sale prices? In chapters 2 and 7, I'll examine the possibility that the test could be a loss leader for a potentially lucrative biobank—particularly because of clauses specifying that the genetic analysis remains the property of the firms. A central National Institute of Health biobank was one of the demands made on government by the private interests in PCAST. But without a nationalized health service to recruit donors free of charge, which has benefited the 500,000-genomes-strong UK Biobank, there's no alternative for U.S. corporations but to recruit privately, as cheaply as possible. Some observers think that the acquisition of health and genetic data, as well as a patent portfolio, is the real business strategy of the retail genetics industry.[76] This supposition began to look highly plausible in June 2012, when 23andMe was awarded a potentially profitable patent on a genetic variant that appears to protect against a high-risk mutation for Parkinson's disease.

Those customers who buy these tests probably labor under the illusion that they continue to own their personal data and their tissue samples. That argument was put forward in a recent blog debate by neoliberal proponents of retail genetics who oppose FDA regulation of the tests: that it's not up to government to tell people what they can do with their own bodies and with information about their bodies.[77] But that argument is mistaken: as I'll elaborate in chapter 2, once tissue has left the body, common law traditionally treats it as *res nullius*—no one's thing. Originally that tissue was presumed to have no value because it was diseased. But modern biotechnology has radically altered the financial position, although the legal position is largely unchanged.[78] That's left a vacuum for corporations, researchers, and universities to claim legal ownership of the tissue once it's in their hands.[79]

Precisely because so few consumers realize that they're actually surrendering ownership of their tissue to the firm once they've sent off the sample, and because the corporate interests in retail genetics are often powerful, consumers are vulnerable. Me Medicine is typically portrayed as empowering, but the real power and legal rights rest with the corporate interests in this case, particularly when they're backed up by neoliberal government policy. Those same policies increasingly spell trouble for We Medicine policies in public health and funding for medicine outside the private sector, as the grim prospect of austerity measures in the public sector stretches out into the foreseeable future.

In addition, Me Medicine could well increase the demands by patients on healthcare systems, for example, if they've bought retail genetic tests that reveal false positives seemingly requiring treatment when they're actually perfectly healthy. (Think back to David Ewing Duncan's urgent phone call from Kari Stephansson of deCODE Genetics, insisting that Duncan should go onto statins immediately because of results that were later contradicted by other DTC services.) Such extra demands on healthcare services are a sort of externality: the costs are passed on to public or private insurance systems. Whether it's a public or a private provider who picks up the tab, in neither case are they borne by the direct-to-consumer test provider.

Risk sharing is the principle behind both insurance-based U.S. health care and UK semisocialized medicine—even if the groups across which the risk is shared may differ. That principle is threatened when risk stops being shared because low-risk individuals identify themselves as such through personal genetic tests and high-risk patients are either booted out of the scheme altogether or limited to a very minimal package of health options. So the conflict between neoliberal ideology and social solidarity, Me and We, is central here and in many other areas. In chapter 7, I'll provide a much more extended discussion of the notion of the public commons. Now, however, I want to move on to the final hypothesis about why Me Medicine is on the rise: the elevated status in our culture of choice and autonomy.

THE SACREDNESS OF PERSONAL CHOICE AND INDIVIDUALISM

We've already seen that personalized genetic testing plays heavily on the first-person singular: deCODEme, Knome, 23andMe, and their like. That's the personal part: the choice part is equally important. Autonomy and its partner, choice, are the paramount values in the dominant paradigm of medical ethics[80] and, arguably, in society as a whole.

In medical ethics, autonomy originally played an important role in elevating patient-centered care over medical paternalism, which is the notion that "doctor knows best" in judging the interests of the patient. The ideal of patient-centered medicine insisted instead that the wishes of competent adult patients should be respected, even if it meant refusing potentially life-saving treatment. Autonomy is also central to the Declaration of Helsinki principles for research ethics, to protect those who might be coerced into consenting to take part in trials. Autonomy came to be seen as the most important of the "four principles" (along with beneficence, nonmaleficence, and justice), in the approach that dominated the teaching of medical ethics for many years in the United States and United Kingdom.[81] Outside observers, however, have contended that the overemphasis on autonomy has impoverished medical ethics as a whole.[82]

Yet the supremacy of autonomy and individual choice in medical ethics hasn't gone unchallenged.[83] "The choice model falsely reduces all ethics to whether something is genuinely chosen, which results in minimising all other injustices."[84] Feminist bioethicists have asked whether autonomy is too individualistic and if it needs to be balanced with a focus on relationships and power.[85] Some medical ethicists have examined the comparative claims of autonomy and trust[86] or have advocated a more communitarian approach.[87] Others have followed such thinkers as Hans Jonas in arguing that in an age of unpredictable technological change, we need to think more about our communal responsibilities than our individual rights.[88]

More sophisticated concepts of autonomy do distinguish between acting on your immediate inclination and acting in accordance with your stable value system, arguing that only the second kind of choice is genuinely autonomous.[89] But that seems a long way from the manner in which personal choice is used as a mantra in personalized medicine, as later chapters will show.

The exalted place of personal choice is not a cultural universal. In France, for example, the values of solidarity and protection for the vulnerable regularly trump free markets, choice, and individualism in framing bioethics laws.[90] Likewise, the Nordic countries are concerned that their more communal values may be threatened by an overemphasis on choice in consumer medicine.[91] But in the United States, it's been said, liberals are almost as prone as conservatives to elevate individual freedom over the welfare of society.[92] By selecting the "right to choose" as likely to be the most psychologically and politically effective counterweight to the "right to life" in the abortion debates, progressives unwittingly hitched their wagon to what later turned out to be the ubiquitous neoliberal ideology of choice.

It's much the same with commercial "surrogacy" (more accurately termed pregnancy outsourcing, since the "surrogate" mother is the real mother in our common-law system). There, paternalism—denying someone freedom of choice—is used as a knockdown argument against which there's meant to be no recourse: "That doctors would be so paternalistic as to deny women the option of using a surrogate if the surrogate were willing to do so is simply outrageous."[93] The same tactic is used with organ sale: "To ban a market in organs is, paradoxically, to constrain what people can do with their own lives."[94] But some see this maneuver as a form of censorship—and censorship is not known, of course, for enhancing individual choice:

> [This] argument shows why focusing only on autonomy silences other ethical concerns, as to deny the validity of choice or the permissibility of a chosen act is to be "paternalistic," "disempowering," "moralistic," "patronising"— and lots of other not so nice things: . . . paternalism is a particularly dirty word in ethics. As a result it becomes impossible to critique any practice if someone—anyone—has chosen it—as to do this is apparently to deny and undermine someone's autonomy. . . . In this way then the consent model reduces all ethics to choice and silences and trumps other ethical concerns. This does not protect the individual, but leaves him or her vulnerable and open to exploitation.[95]

In a less blatant manner—but probably only because the debate hasn't really got going yet—discussion about direct-to-consumer genetic testing has centered on whether it enhances personal responsibility for detecting and directing your own future health or whether the information available to consumers is too misleading to allow a genuinely informed choice. Those questions matter, but they aren't the end of the affair. In particular, they have nothing whatsoever to say about the harmful effects of Me Medicine on We Medicine, particularly when denial of free choice is used summarily to dismiss vital public health measures such as vaccination programs or travel restrictions during epidemics.[96]

Choice isn't a knockdown argument in personalized medicine. As with prostitution or pregnancy outsourcing, even if individuals make choices, those choices influence and are influenced by the social context in which the practice is embedded. It is a blatantly false assumption that whatever you do, you've chosen to do—and that you've made your individual choice independently of any social, political, or economic factors. That's actually a very simple point, but it has to be made and repeated constantly in our culture. In the unfamiliar context of a new biotechnology such as retail genetics, there's a particular temptation not to think through the wider social consequences but simply to fall back on the old familiar argument: "what's the problem, if that's what people choose to do?"

The philosopher Zahra Meghani argues that we always have to understand individual medical choices, such as whether to go abroad to buy eggs or to hire a "surrogate" mother, in the context of global neoliberalism and its core policies: privatization, deregulation, and commodification.[97] Rather than an apolitical, one-size-fits-all argument like choice, we also have to understand local realities. Of course, it's not just the Third World that possesses its own local realities: very particular factors characterize American culture as well. In this chapter, I've made a start on examining some of those factors that might be particularly relevant to the push for consumer medicine—including a sense of threat, consumerist narcissism, and corporate interests. Taking personal choice at face value closes down that analysis before it's even properly begun: it's a lazy argument that does none of the necessary work.

The effect of the failure by those on the political left to challenge the mantra of personal choice is that progressives have too readily retreated from challenging the neoliberal deregulation of biotechnology and corporate interests when representatives of those interests accuse them of wanting to limit consumers' freedom of choice.[98] They haven't done all they could to identify the phenomenon of Me Medicine and to challenge its reliance on personal choice as a knockdown argument. Simultaneously, progressives have found it difficult to challenge the reaction against communitarian forms of medicine as wrong because they limit individual choice—the argument used to lambast President Obama's healthcare plans, vaccination programs, or swine flu epidemic restrictions.

The situation is worsened in the United States by the "stem cell wars," in which it was assumed that progressives would automatically be on the side of science, standing against the evangelical right's campaign to outlaw embryonic stem cell research. Liberals and progressives may be tempted to ignore moral issues in the new biotechnologies because they fear being lumped in together with the religious right. In the case of new biotechnologies such as direct-to-consumer genetic testing and enhancement, the corollary is that they may be unwittingly prone to support Me Medicine against We Medicine, even when their sympathies would more naturally lie with the latter. These issues are nothing if not complicated: I'll examine a particularly tough one for political progressives in chapter 3: the development of supposedly "race-specific" drugs such as BiDil.

While the notion of the social contract may still be more or less intact in Scandinavia, it was never particularly strong in the United States, and even among academics and activists it's been weakened—inadvertently—by well-grounded liberal critiques of the way in which it favors one sex or race over another. The social contract as an instrument of civic subordination was brilliantly analyzed by Carole Pateman in her 1988 book *The Sexual Contract* and by Charles Mills in his 1997 work *The Racial Contract*. Pateman's crucial insight was that liberal contractarian theory is blind to the way in which the "original position,"[99] from which the state is constructed by voluntary contract, is not really "original" at all. It must be preceded by another sort of compact, in which male domination over women has been established through the mechanism of the patriarchal family, since those establishing the contract in the "state of nature" are generally assumed to be men. Mills builds on this insight to demonstrate how even after the abolition of slavery, people of color likewise continue to be subordinated and oppressed through the mechanism of a supposedly consensual contract in liberal democracy: the consent of the governed.

In their collaboration *Contract and Domination*, Pateman and Mills differ crucially on how refractory the concept of contract really is. While Mills thinks that contract theory can be "modified and used for emancipatory purposes,"[100] Pateman continues to maintain that contract is inherently an instrument of domination—although other feminists have argued that what's wrong with the sexual contract is not that it is a contract but that it is sexual.[101] Because the so-called marriage contract—actually not a legal contract at all—was blatantly oppressive, feminists had good reason to distrust the social contract more generally. But along with the other factors sketched out in this chapter, that skepticism may have inadvertently encouraged distrust of "We"-ness when it represents false inclusivity.

Unconstrained commodification of the body seems to Pateman to make the concept of the social contract even more suspect. "Commodification is proceeding at such an extraordinarily rapid rate; there is virtually nothing left now that is outside the reach of private property, contract and alienation," she remarks to Mills in a dialogue at the outset of their joint book. "That is one reason why I'm much less happy than you with trying to salvage contract theory."[102] But without some notion of common interests in healthcare embodied in something like contractual form, what protects us against the unstoppable rise of Me Medicine? That may sound like pastry in the stratosphere, but international agreements and treaties such as the European Patent Convention have been used to good effect in protecting the human genome, as the joint property of humanity, against "the great genome grab" of commercial patents. Likewise, Article 14 of the 2005 UNESCO Universal Declaration on Bioethics and Human Rights introduces a principle of social responsibility for health, transcending shopworn individualistic bioethics.

I'll return to these considerations in the final chapter, when I try to establish how we can reverse the trend elevating "Me" above "We" in how we use modern biotechnology—how we can reclaim it for the common good. Now it's time to analyze that technology in greater detail. In suggesting four possible reasons why "Me" is privileged over "We"—threat, narcissism, corporate interests, and the sacredness of choice—I've begun by situating the technology in its wider cultural and political context. The next step is to apply those four hypotheses to four areas of Me Medicine: retail genetics, pharmacogenomics, private umbilical cord blood banking, and enhancement technologies.

2

"YOUR GENETIC INFORMATION SHOULD BE CONTROLLED BY YOU"

Personalized Genetic Testing

IN JUNE 2011, AT THE THIRD ANNUAL CONSUMER GENETICS SHOW, the biotechnology company Illumina Incorporated unveiled its *MiGenome* application for the iPad tablet computer. (That's a double dose of the first-person singular: the possessive "my" added to the "I" in iPad.) Once you'd had your entire genome sequenced by Illumina—at a price of $9,500, reduced from the previous $19,500—*MiGenome* would ostensibly allow you to check your susceptibility to genetically based disorders. You could also find out how, given your genetic makeup, you would probably respond to particular drugs. *MiGenome* would display your entire genome, but if even the reduced price was too much for your pocketbook, you could buy a cheaper testing package, which would provide the results for a more limited range of genetic markers. As we discussed in the last chapter, like *MiGenome*, these "retail genetics" tests are steeped in the "Me" brand, with company names such as 23andMe, deCODEme, and MyGenome.[1]

Even when government has tried to regulate the consumer genetics sector, it has unwittingly accepted the language and underpinning philosophy of Me Medicine. In Massachusetts and Vermont, for example, proposed genetic privacy legislation declares, in identical language, that "genetic information [is] the exclusive property of the individual from whom the information is obtained."[2] But our common law traditionally has held that we have no property in tissue once it's left the body—whether through the minor inconvenience of a "spit kit" or a major surgical procedure. It was considered *res nullius*—no one's thing—because it was presumed to be diseased and to have no commercial value. So, no, you don't necessarily own your tissue or the information derived from it—still less own it exclusively. Whether it would be a good thing if you did will be discussed in a later section of this chapter—but the Massachusetts and Vermont legislators seemed unaware that you don't. The language of "I, me, mine" is by no means straightforwardly appropriate to genetic information, common though it is.

Yet oddly enough, the Vermont bill also stipulates that genomic information should be part and parcel of *We* Medicine. Elsewhere, in section 9336(e), the proposed statute states:

> Information derived from the sequence of the human genome shall be part of the public domain and shall not be considered the property of any individual. Nothing in this chapter shall be considered to grant an ownership right to any individual or entity utilizing the publicly held information from the sequence of the human genome in the furtherance of a venture or enterprise, including any genetic goods, products, or services.[3]

What could explain such a blatant contradiction? Possibly the legislators were trying to distinguish between *the* human genome, conceived as the common heritage of humanity,[4] and *a* human genome, exclusive to one individual. Or perhaps they were grappling, without fully realizing it, with the intricate and genuine conflict between Me and We Medicine, as emblematized by direct-to-consumer (DTC) personalized genetic testing. Their concern in the second quotation is "the sequence of the human genome in the furtherance of a venture or enterprise, including any genetic goods, products or service"— and the most visible and contentious of those services, at present, is DTC genetic testing. Another New England state, Connecticut, already prohibits it, requiring a doctor to be the intermediary between test and patient.

Consumerized genetic testing has become a lightning rod for controversy, both because it involves a direct link between researcher, industry, and consumer and because it is predicated on premises that genomic research does not fully support. Furthermore, different DTC companies offer different results for identical DNA samples. It's not yet possible to aggregate independent risk factors into a net risk score.[5] Reliability has not been certified, and no professional organization standardizes the tests.

Although it's still not a huge sector in terms of dollars or customers, retail genetics has become a highly visible symbol for personalized medicine more generally. Recently it has also become the site of conflict between those who think that buying into DTC genetic testing should be a matter of personal choice and those who believe that it needs public regulation. About half of U.S. states either prohibit or limit it, like Connecticut, but the rest leave it alone. So there's a clash of Me and We ideologies, with proponents of direct-to-consumer genetic testing using the individualistic language of empowerment, choice, and responsibility against the notion that society has an interest in limiting potential risks.

As the 23andMe website says, "Take charge of your health and wellness: let your DNA help you plan for the important things in life." The firm also declares: "We believe your genetic information should be controlled by you."[6] A former direct-to-consumer genetic testing firm, Navigenics, stated: "We use the latest science and technology to give you a view into your DNA, revealing your genetic predisposition for important health conditions and empowering you with knowledge to help you take control of your health future."

At first glance, that mission seems laudable. Particularly when DTC genetic testing is targeted not at comparatively trivial traits such as athletic ability or earwax buildup—and yes, you can get tests revealing your supposed personal predisposition to both of those—it could be seen as harmless at worst and even admirable in more serious cases to confront your genetic risks. That seems particularly plausible for tests of your risk of passing on inherited diseases to your offspring, rather than your own susceptibility. In fact, some bioethicists argue that parents at risk for possibly life-threatening conditions have a moral duty to undergo genetic testing in the form of preimplantation genetic diagnosis of the embryo—a responsibility that they claim could and should even become a legal requirement.[7]

In January 2010, a company called Counsyl began offering a $349 saliva sample test to identify alleles for common but serious hereditary illnesses such as cystic fibrosis and sickle-cell disease. For those and other recessive genetic conditions, prospective parents can be carriers without manifesting the disease themselves. Having yourself tested seems quite a responsible thing to do before you start a family—intrinsically recognizing "We"-ness with your potential child.

So is it anything more than trivially symbolic that retail genetics uses the language of "I, me, and mine" so readily? To begin analyzing these questions, we first need to get a better understanding of the historical background of personalized genetic testing.

"THE REST OF THE POPULATION CAN BE ALLOWED TO PUFF AWAY CONTENTEDLY"

The announcement, on June 26, 2000, of the draft sequencing of the entire human genome generated tremendous optimism and excitement, which still linger in the aura surrounding personal genetic testing. In chairing a joint announcement with Prime Minister Tony Blair, President Bill Clinton called the announcement of the Human Genome Project's draft results "a day for the ages."[8] Clinton went on to declare, "This landmark achievement will lead to a new era of molecular medicine, an era that will bring new ways to prevent, diagnose and treat disease."[9]

It was widely expected that common diseases would be found to have a considerable genetic component and that once the genetic code was cracked, clinical cures would quickly result. In 2001, the Human Genome Project's director Francis Collins and his colleague V. A. McKusick predicted that there would soon be reliable genetic tests for up to a dozen common conditions, so that general practitioners would essentially become genetic counselors.[10]

Two years later, there were a dozen private companies advertising genetic susceptibility testing for individuals to purchase via the Internet, but very little progress had been made in tracking the genetic causes of the most widespread forms of illness. By 2009, the number of these testing firms had risen to thirty, some offering very specialized testing for characteristics such as athletic ability—blazoned under the advertising slogan "Olympic success might be in your future!"[11] In 2011, the consumer genetics industry was estimated to include nearly one hundred companies.[12] These firms' strategy has rested not so much on the very patchy medical and scientific evidence base as on convincing venture capital that a demand for personalized testing exists, thereby creating high expectations for returns—and then creating the demand from consumers to match.[13]

Compared to the logarithmic increase in the number of retail genetics companies, there's been nothing like the same exponential rate of growth in genetic medical diagnostics, let alone cures.[14] As one journalist put it: "After 10 years of effort, geneticists are almost back to square one in knowing where to look for the roots of common disease."[15] For example, in the major area of cardiovascular illness, a twelve-year study of 19,000 women found no significant correlation between 101 genetic variants linked to heart disease in genome-scanning studies and the actual incidence of such disease.[16] This study typifies the position for common illnesses caused by many genes—but not by genes alone. Unfortunately, those common diseases are the biggest killers.

There's been a great deal of discussion about why the Human Genome Project seems not to have fulfilled its scientific promise. One major surprise was that the human genome contained far fewer genes than expected: between 23,000 and 25,000, rather than the predictions of as few as 50,000 and as many as 140,000. (You may recall a certain amount of shamefaced speciesism at the time, comparing our paltry number of genes to those of fruit flies and other so-called primitive creatures.) But you might think that fewer genes would mean simpler diagnostics and less complicated pathways to therapies, so that if anything, Collins and McKusick's prediction would have been too modest.

Instead, subsequent developments showed that our comparatively limited palette of genes can create complex color shadings. Variation is hidden in a diversity of nonlinear interactions between the proteins coded by genes[17] and between genes and environmental factors affecting the way they're expressed.[18] That second set of factors, studied by the emerging science of *epigenetics*, involves modifications to our genetic material affecting the ways genes are switched on or off. While each of our cells contains the same genetic "code," that code can produce everything from eyeballs to teeth. We're really only just beginning to unravel the full complexity of the epigenetic modifications that ensure that each cell does what it's supposed to.

Already, however, there's a radical change of mood in genomic science— and "science is just as prone to mood swings and fashions as any other human activity." Although "there was a period when the prevailing orthodoxy seemed to be that the only thing that mattered was our DNA script, our genetic inheritance . . . that can't be the case. . . . The field is now possibly at risk of swinging a bit too far in the opposite direction, with hardline epigeneticists almost minimizing the significance of the DNA code. The truth is, of course, somewhere in between."[19]

Even at the time the draft human genome sequence was announced, it was already known that there were no "good" or "bad" genes, just complex relations between networks of genetic and epigenetic factors.[20] Only for a very small minority of illnesses is there a direct one-to-one correlation between having a particular form of a gene and manifesting a disease.

Single-Gene Disorders and Mendelian Genetics

The Mendelian genetics that you learned in Biology 101 is based on experiments done in the 1860s with plant characteristics whose variations are caused by single genes: in the peas with which Mendel worked, for example, the variations included wrinkled versus round seeds, tall versus dwarf stems, or grey versus white seed coats. In each case, one characteristic is dominant and the other recessive, with the phenotype (the physical appearance) determined by the dominant allele in the genotype (the genetic makeup of the plant). If, for example, a pea plant has one allele for wrinkled seeds and one for round seeds, the seeds in the pods will be round, with the recessive wrinkled allele suppressed by the dominant round one. But if crossed with another pea plant that also has one wrinkled and one round allele, there's a one-in-four chance that the new offspring plants will inherit two recessive wrinkled versions of the gene and display the wrinkled seed form.

In a very few human medical conditions, Mendelian genetics does apply straightforwardly: for example, in Huntington's disease, which is a *dominant* condition linked to a single gene. Inheriting one unfavorable allele of that gene from *either* parent is sufficient for the disease to be manifested, because the disease-producing allele is dominant over the "healthy" variant from the other parent. By contrast, in *recessive* conditions such as cystic fibrosis, the child must inherit two disease-linked alleles, *one from each parent*, before the condition manifests itself. If only one parent conveys the unfavorable allele to the embryo, the dominant "healthy" allele from the other parent effectively cancels it out. Mendelian genetics also applies to beta-thalassemia, sickle cell disease, and Canavan's disease, but comparatively few illnesses are purely "in the genes." Even for Huntington's disease, things aren't that simple: the age of onset depends on the number of repetitions of the genetic marker.

More typically, the way in which common diseases are linked to tens or even hundreds of genes, each explaining only a tiny fraction of the variance between healthy and sick individuals, means that genetically based diagnosis in any particular case is at best a probability rather than a certainty. Family history is still a better predictor than genetic analysis of common conditions like cardiovascular disease.[21]

There's also a subtle but crucial difference between predicting the likelihood of a particular *individual* contracting a particular disease and testing entire *populations* for genetic susceptibility.[22] This distinction—another form of Me versus We—is highly relevant to personalized direct-to-consumer test-ing. Although the DTC companies assert that their tests are not the equiva-lent of a doctor's individual diagnosis, one study found that a third of custom-ers believed that they were in fact purchasing a diagnosis.[23] Yet the results are not individual diagnoses but rather comparisons of how any individual stands in comparison with general populations when it comes to the likelihood of contracting a particular disease.

So how could the Human Genome Project scientists have believed that the majority of common diseases would follow anything like the one-to-one Men-delian pattern? If they themselves knew better, why did they allow the im-pression to be given that things would be much simpler than they've turned out to be? Teasing out the answer requires us to look at the history of the project—in particular, at the possible tension between the massive amounts of public money that poured into it and the private interests of commercial firms, particularly tobacco companies.

The Human Genome Project (HGP) appears to be a genuine exemplar of We Medicine, at least in its original conception—regardless of whether some of its applications were later converted by commercial genetic testing compa-nies into Me Medicine. It was certainly public in its original funding, through the U.S. National Institutes of Health, UK charity the Wellcome Trust, and the UK Medical Research Council. By releasing into the public domain some 1.8 million genetic markers called SNPs (single nucleotide polymorphisms, explained at greater length in the next section of this chapter), the publicly funded HGP also laid the basis for the private genetic testing companies to market their SNP-based wares.[24]

But some researchers have asked whether the HGP's grand aims for hu-mankind were in fact dictated by private commercial interests from the very start. Those interests, they allege, were always focused on Me rather than We— using genome research to identify *individuals* who were particularly suscepti-ble to lung cancer if they smoked and not on identifying *population* propensities to a much greater range of illnesses. What's the evidence for this astounding claim?

As early as the 1950s, the tobacco industry began to promote the idea that an unknown gene both drove people to smoke *and* predisposed them to lung cancer. It was in the industry's interest, some researchers allege, to promote genetic screening in order to convince ambivalent smokers that they had no need to quit.[25] In the words of a memorandum sent by the public relations firm Burson-Marsteller to the tobacco firm Philip Morris: "A simple test might eventually be devised to tell a smoker whether or not he is at risk. This would put the burden for any consequence from smoking on the individual and would clear the way for the non-susceptible population to smoke with a clear conscience."[26]

Then, as the industry consultant Frank Roe put it: "the rest of the population can be allowed to puff away contentedly and without serious risk."[27] This strategy clearly puts the burden of responsibility entirely on the individual. We Medicine measures such as public campaigns to raise awareness of the fatal risks of smoking would be pointless if those members of the general public who were not genetically susceptible ran no risk.

That single memorandum wasn't just an isolated occurrence. The industry's Council for Tobacco Research gave evidence to the U.S. House of Representatives in 1994 that it had already awarded nearly $225 million to sponsor "pioneering work in identifying familial cancers, the role of genetic factors in cancer formation and the identification of oncogenes [cancer-related genes]."[28] Scientists funded by the tobacco industry allegedly published spurious findings to convince funders that human genome sequencing would be useful in predicting who develops common diseases.[29]

No one claims that tobacco-funded scientists *deliberately* falsified results; rather, "the false claims resulted from poor science and a process by which tobacco-funded scientists benefited from fast-tracked careers, financial and political support, and access to the media to promote the industry's messages: that cancer is a genetic disease and prevention depends on screening people's genomes so that lifestyle and medical advice can be targeted at those at high genetic risk."[30] Even when the research wasn't spurious, it was orchestrated with a view to determining one-to-one individual susceptibility of the Mendelian kind. It's no surprise, if this is true, that human genome sequencing hasn't turned out to be as useful in unraveling common diseases as the grander claims suggested it would be, since common diseases rarely fit the simple Mendelian model.

These are unsettling charges, but they have been made by several separate groups of researchers, are backed up by extensive data gathering, and draw from damaging documents revealed by the tobacco firms only after litigation forced their hand. These sources have revealed secret meetings between leading scientists and British American Tobacco as early as 1988, before the launch of the HGP. Genewatch UK researchers also claim to have uncovered links between the tobacco industry and Kari Stefansson, president of one of the original DTC companies, deCODE, now part of U.S. biotech firm Amgen.

Meanwhile, work published by the Mayo Foundation for Medical Education and Research has drawn attention to the consistency of the tobacco industry's strategy, from pre-HGP days to the present.[31]

Although the genetic basis for tobacco addiction may originally have looked like an industry pipedream—excuse the pun—it now appears more scientifically plausible. In the past, however, the industry sought to *deny* that nicotine was addictive. Now their strategy, according to the Mayo researchers, is to accept that smoking is indeed addictive but to subvert genetic research into narrowing the risk of addiction to "unsafe" smokers.

> The search for a genetic basis for smoking is consistent with industry's de-
> cades-long plan to deflect responsibility away from the tobacco companies
> and onto individuals' genetic constitutions. Internal documents reveal
> long-standing support for genetic research as a strategy to relieve the to-
> bacco industry of its legal responsibility for tobacco-related disease. Indus-
> try may turn the findings of genetics to its own ends, changing strategy
> from creating a "safe" cigarette to defining a "safe" smoker.[32]

This quotation indicates that the history behind the Human Genome Project and the search for an individualized model of responsibility for health isn't *just* history. That's important because it deflects the charge that all this is merely of interest to historians of medicine. But even if we accepted for argument's sake that the motivating forces behind personalized genetics were tainted, that wouldn't allow us to conclude that the science is necessarily faulty. To evaluate that question, we need to look at a different set of evidence.

EVALUATING PERSONALIZED GENETIC TESTING

The testing package offered by Illumina would sequence your entire genome: that is, it would determine the order of every single one of your base pairs, resulting in a library or personal genomic database of three billion letter combinations and potentially laying the basis for genuinely personalized medicine.[33] Whole-genome sequencing is what the Human Genome Project achieved for the first time, with final results announced in 2003, following the highly publicized launch of the draft sequence in 2000. But those results weren't personalized; that is, no single individual was sequenced. The sequenced genome was a composite of several anonymous people recruited through advertisements in the *Buffalo News*.

Mapping this composite genome required $3 billion in funding and thirteen years of research. Since then, the cost of whole-genome sequencing has tumbled so fast, assisted by $14 million in grants for that purpose during 2011 from the National Human Genome Research Institute, that the thousand-dollar genome is now here, as was announced in January 2012. Going that figure one decimal point better, a report from the JASON group of science advisers at the nonprofit Mitre Corporation predicted that "the $100 genome is nearly upon us."[34]

Whole-genome analysis takes the raw data obtained from sequencing and applies filters that target certain parts of the genome, dimming the "noise" of less relevant data. Since the genomic scientist and entrepreneur Craig Venter published the complete sequence of his own genome in 2007, a number of celebrities, from Glenn Close to James Watson (of double helix fame), have had their entire genomes sequenced. Meanwhile, the Personal Genome Project run by George Church at Harvard has enrolled a growing number of subjects—with sixty-four genomes completed at the time of writing—for whole-genome analysis, the first guinea pig being Church himself.[35] (Although itself not for profit, the Personal Genome Project has a link to Google Health for phenotype collection, and Google in turn has a potential link to 23andMe through its cofounder Sergei Brin, who is married to 23andMe's CEO, Anne Wojcicki.)

Comparatively cheap whole-genome analysis may perhaps become standard in the future—or it may not. We saw that Illumina charged nearly ten times as much as a "thousand-dollar genome," even though the firm has reduced its prices. Yet a thousand-dollar test is considerably more expensive than the cheapest packages offered by leading DTC companies, which typically range from $399 to $2,000 but can come in as low as $99—or in some cases, free, provided that the genetic data is treated as the property of the firm and that customers agree to provide additional information about their health. On August 1, 2011, under the slogan "Roots Into the Future," 23andMe announced free testing for ten thousand African American customers, on condition that the data and DNA samples remain in the company's hands.

Even when they don't offer their services free of charge, how do these companies achieve their bargain-basement prices? Essentially, they don't sequence the entire genome: they concentrate on a far smaller number of markers—up to 1.8 million (that may seem like a lot, but it's still a far cry from three billion). This strategy brings down the price, but it has also been criticized for diminishing the scientific value of the enterprise. Consumer genetics firms all choose to sequence different sets of markers, and thus they return very disparate results.[36] Researchers from the Erasmus University Medical Center in the Netherlands and Harvard Medical School examined two widely available tests and found that their results for common diseases such as diabetes and prostate cancer were radically inconsistent because of the low numbers of markers involved and the small amount of genetic variance that each marker explains.[37]

What Is an SNP?

The main target of DTC genetic testing is the single nucleotide polymorphism (SNP), a point where the genomes of different individuals vary by a single DNA base pair.[38] These markers are derived from genome-wide association studies (GWAs), population-level genomic research aimed at determining the correlation between common diseases and certain areas of the human genome. Researchers compare people with the disease and similar people without the illness, obtaining DNA from each participant and placing it on sequencing chips, which analyze the person's genome for strategically selected markers of genetic variation, that is, SNPs.[39] They then determine whether particular SNPs are associated with particular diseases by statistical significance techniques, converting the numbers into disease susceptibility risk figures.

Clearly, a higher level of significance is required to establish a definite disease susceptibility association if the number of SNPs is limited, since each SNP only covers a fractional amount of association with any of the common diseases. But commercial DTC firms must of necessity limit the number of SNPs that they examine. Of the leading retail genetics companies, 23andMe covers about 600,000 SNPs, and deCODEme, roughly 1.2 million. Navigenics had tested for about 1.8 million markers, of which 906,000 were SNPs and the remainder probes for copy number variation.[40]

Yet even Navigenics tests were found in a study led by the founding director of the Cleveland Clinic, Dr. Charis Eng, to provide imperfect guidance on their own—without family history–based assessment—for any individual's personal risk of developing three common forms of cancer (breast, prostate, or colon).[41] While family history indicated that eight individuals were at high risk for breast cancer, only one of the eight was classified as high risk when assessed via personal genetic testing. Overall, family history assigned twenty-two individuals to the hereditary elevated risk category, but DTC testing identified only one of these individuals as high risk. The researchers also assessed nine individuals with hereditary risk for colorectal cancer, five of whom had proven mutations defining inherited colorectal cancer syndromes. None of the nine was classified as high risk when assessed through DTC analysis.

Eng and her colleagues are convinced that family health history is still the gold standard in personal disease risk assessment—and it's cheaper too. She adds that this type of information can be readily gathered by the patient—which might actually enhance your autonomy and sense of responsibility for your own health more effectively than paying a corporation for a spit test. (In fact, taking your own family history could combine "Me" and "We" quite nicely.)

As Martin Richards has written of his own experience of DTC testing, "The companies' literature seems to promise that they can tell us more about ourselves than we can know for or by ourselves alone. In that sense they actually undermine our autonomy."[42] Perhaps DTC tests might even decrease your sense of individual control if you find that the results you've paid good money for contradict one another, or if you realize that nothing can be done about the condition to which you've now been told you're susceptible.

There's also a very real risk that asymptomatic healthy people may come to define themselves instead as merely "presymptomatic," making us all patients from the cradle to the grave.[43] As George Church of the Personal Genome Project says, "Even if these highly predictive and actionable [variations] are considered rare, everyone is at risk and should be just as willing to spend on this as on fire insurance and other unlikely contingencies."[44] Church's view illustrates the way in which direct-to-consumer genetic testing plays, whether deliberately or not, on a sense of threat. But how does permanent patienthood enhance our autonomy?

For those who have a family history of risk for genetically linked cancers, it's perhaps a different matter—but not all that different. The most recent policy update from the American Society of Clinical Oncology[45] accepts that genetic testing for personal cancer susceptibility is now a routine part of clinical care, especially for high-penetrance mutations like the alleles of the *BRCA1* and *BRCA2* genes implicated in some breast and ovarian cancers. However, the society also notes that such cancers, though serious, account for only a small percentage of all cancers.

Cancer-related combinations of sequence variants have been identified through genome-wide association studies, along with over one hundred relatively common SNPs linked to parts of the genome associated with cancer in a yet undetermined way. Their penetrance varies with epigenetic, lifestyle, and environmental factors. The American Society of Clinical Oncology believes that testing for these SNPs is of uncertain clinical value, because the risk is generally too small to serve as the basis for clinical decision making.[46] By contrast, a family history of breast and ovarian cancer, for example, would alert a clinician to order a direct and specific test for the *BRCA1* and *BRCA2* genes implicated in some such tumors. (In any case, *BRCA1* and *BRCA2* testing isn't offered by many DTC companies because of restrictive and expensive patent protection on those genes, raising the cost of the tests to as much as $3,500.)[47]

There's also a substantial risk of false positives and false negatives, which is exacerbated by patients' cognitive bias in favor of attributing more certainty to genetic information than the probabilities warrant. Andrew Wilkie, Nuffield Professor of Pathology at the University of Oxford, believes that "Patients and families want answers that give them certainty."[48] But unfortunately, that's not how genetic analysis generally works. Even though there is an unusually strong correlation between the *ApoE4* genetic allele and Alzheimer's disease, for example, 77 percent of people with the unfavorable allele *don't* develop Alzheimer's disease—so identifying the allele implies a false positive for them—while 47 percent of people who *do* manifest Alzheimer's disease don't have the mutation (a false negative). If a genetic test can't confer *absolute* certainty in this case, where the association is unusually strong, then it's all the more unlikely to provide the certainty that patients crave in other situations. "Is this technology just a distraction from focusing on the large preventable environment component?" Wilkie asks, focusing squarely on "Me versus We."

The American Society of Clinical Oncology guidelines recognize that some commentators[49] claim that DTC genetic tests do good by making patients feel in charge of their own health and by motivating them to pursue healthy behaviors. However, the society argues that these untested benefits must be balanced against iatrogenic (doctor-induced) harm and low clinical utility. The oncologists fear that they may be put in a difficult position if they haven't requested the original test, but the patient wants a follow-up. This isn't just a matter of medics protecting their professional position: there are genuine worries about the need for counseling, interpretation, and professional advice when a genetic susceptibility to cancer is revealed.

The society is also concerned about the lack of an evidence base: over 40 percent of the genomic variants used in commercial assays haven't been replicated in meta-analyses involving many studies, which is the gold standard for clinical evidence. No published studies have yet assessed the reliability of the algorithms used by the retail genetics companies to create the risk estimates fed back to patients. "Because these tests have uncertain clinical validity, they are not currently considered part of standard oncology or preventive care," the guidelines conclude.[50]

But will the patient who has paid hundreds or thousands of dollars for a test accept that it has no value beyond the merely "recreational"—and that her doctor is entitled to refuse to act on its findings? Even the genetic testing industry's own newsletter, *Genomics Law Report*, thinks not. It's quietly skeptical of 23andMe's disclaimer that its products are for recreational use only, pointing out quite cannily that consumers won't part with several hundred dollars unless they really think their health will benefit.[51] After all, didn't the 23andMe website urge patients, "Take care of your health and wellness"?

In fact, it's far from certain that the test results actually motivate consumers to make healthy lifestyle changes, as the DTC companies often claim that they do. A study in the *New England Journal of Medicine* of 3,639 retail genetics customers found no significant improvements in their diet or exercise regimen.[52] Likewise, a smaller study involving in-depth interviews with twenty-three "early adopters" revealed that very few intended to make any changes in their lifestyle, whatever the tests showed.[53] It's a well-worn truism that morbidity and premature mortality in the developed world arise in very substantial part from smoking, sedentary behavior, and excessive consumption of food or alcohol. None of us needs an expensive test to tell us how far up we need to pull our socks.

It's the specific commercialized form of direct-to-consumer testing for SNP variation that has aroused the greatest skepticism in the medical world and the most serious concern in regulatory agencies. Although in March 2012 the NIH set up something called the Genetic Testing Registry to improve transparency about genetic tests, it doesn't include direct-to-consumer genetic testing services.[54] That same month, a report from the U.S. Institute of Medicine warned that commercially available genomics tests require much more stringent regulatory oversight and more transparent data sharing, after a pharmacogenomic test for chemotherapy regimes against cancer, reported in and then retracted from the high-status journal *Nature Medicine*, proved to be worthless. The IOM report extended beyond retail genetics, also urging tighter controls for genomic tests ordered in clinics. According to the chair of the IOM report committee, Gilbert Omenn, "Nothing short of patient safety and public trust are at stake."[55]

The Government Accountability Office pulled no punches when it published its report from a four-year investigation of ten tests from four retail genetics companies under the headline "Direct-to-Consumer Genetic Tests: Misleading Test Results Are Further Complicated by Deceptive Marketing and Other Questionable Practices."[56] In that same year, the Food and Drug Administration did act speedily to prevent Walgreens from stocking over-the-counter "Insight" retail genetics kits from Pathway Genomics, while maintaining an ongoing investigation into the companies more generally. But Jeremy Gruber of the Council for Responsible Genetics complains that for the most part, "There has been an abdication of leadership in overseeing genetic tests," which "puts a level of uncertainty into the discourse that neither benefits the industry, researchers or consumers."[57]

Given such strong warnings, why is personal genetic testing still being touted as the epitome of personalized medicine? And why does personalized medicine continue to be endorsed by such prestigious figures as Margaret Hamburg of the FDA and Francis Collins of the NIH?[58] Although the IOM report does give cause for concern, we shouldn't fall prey to premature judgment against the whole of personalized medicine because of the scientific and medical failings of some forms of genetic testing, particularly retail genetics. That judgment remains to be drawn, on the additional basis of the rest of the evidence in this book.

Nevertheless, insofar as DTC genetic testing is the self-proclaimed vanguard of the personalized medicine movement, caution is in order—particularly when leading proponents of personalized medicine rest much of their case on the validity of DTC tests. After describing his own experience of retail genetics, Collins writes: "This is a book about hope, not hype. . . . If you are interested in living life to the fullest, it is time to harness your double helix for health and learn what this paradigm shift is all about."[59]

The evidence base, however, simply doesn't bear out Collins's contention that consumerized genetic testing is about hope and not hype. That's why many government and professional bodies want to limit or even ban direct-to-consumer genetic testing, for sound scientific and clinical reasons. In the United States, the numbers are split: twenty-five states allow DTC tests with no restriction, thirteen prohibit them altogether, and twelve only allow some types of tests or require a doctor's involvement.[60] In 2008, California served cease-and-desist orders on twelve DTC genetic testing companies, including 23andMe and Navigenics—requiring them to obtain licenses like traditional clinical laboratories. New York did the same, with the result that some companies ceased their operations in the state altogether.[61]

Those who favor controls, however, often come up against this attitude: "What I do with my genome is nobody's business but my own. Even if I've made a bad buy in purchasing three DTC test kits that return wildly disparate results, is that any worse than buying a used car that turns out to be reluctant to start on a frosty morning?" No legislator would ban the sale of used cars altogether just because some used cars are lemons, this argument runs, although deceptive marketing can be regulated. Why should a consumer's decision to buy into retail genomics be treated any differently? Isn't that excessively paternalistic?

That free-market reasoning seems to lie behind the decisions by the UK's Human Genetics Commission and its Nuffield Council on Bioethics not to prohibit DTC genetic tests altogether or even to regulate them more closely— only to advocate voluntary self-regulation by the industry. The Human Genetics Commission, an advisory body (since disbanded) that included industry representatives, produced nothing more than recommendations of a "common framework of principles" for consumer genetic testing services— guidelines called insufficient even by the *Lancet*, a journal whose political leanings are hardly the equivalent of *Mother Jones*'s.[62] At a public meeting of the HGC in London that I attended on May 12, 2009, one commission member argued that it would be wrong to require evidence of medical need or medical utility for DTC tests when other forms of information available to the public—particularly religious proselytizing—don't have to meet any such test of truthfulness.

Yet buried in an appendix—not highlighted in the summary or main text—the Nuffield Council's report admits that a "strong majority" of respondents to the consultation wanted more regulation of private genetic testing services. The council report also recognizes that other European countries— France, Germany, Austria, and Switzerland—ban DTC testing altogether and that a 2009 survey by the consumer magazine *Which?* showed that four out of five UK readers want retail genetics to be strictly regulated. Nevertheless, the report blandly recommended that "[DTC] companies should voluntarily adopt good practice."[63]

I think that this solution is quite insufficient and that the paternalism charge is actually back-to-front. As I said in my own response to the Nuffield

Council consultation: what people actually want is regulation, which turns the usual argument about paternalism on its head. If it is paternalistic to deny people what they want, then the genuinely paternalistic course is to allow un-regulated genetic testing, not to prevent people from getting unregulated genetic tests on spurious grounds of individuals' "right to know."

WHO OWNS THE GENOME?

It's quite misleading to liken buying a genetic test to any other consumer decision, because our bodies and our genomes aren't just consumer items. In our traditional common law, tissue donors, presumably including people who send a spit sample or cheek swab in for DTC analysis, don't have any ongoing property rights in that tissue. I may own my car or house, but I don't own my tissue once it's taken from my body. Because of this gap in the law, and because people are unaware that there is a gap in the first place, DTC genetic testing companies have been able to follow the lead of other biotechnology corporations and researchers in amassing exclusive rights in tissue held in "biobanks." No DTC customer has yet challenged that position, but if they did, the common-law position would probably be upheld against them.

Biobanks, cell lines, and patents can represent tremendous value: the principal asset in the portfolio of many biotechnology companies lies in the "promissory capital" stored in their patents and databases.[64] (For example, according to a company statement for the fiscal third quarter ended March 31, 2012, 81 percent of total revenue for the biotechnology firm Myriad Genetics was accounted for by tests it offers on the breast-cancer-related genes *BRCA1* and *BRCA2*, for which it holds monopoly patents.)[65] The commercial value of biobanks, cell lines, and genetic patents is becoming better known now, following on from the bestselling story of how tissue from a terminally ill African American woman, Henrietta Lacks, became the source of the multi-billion-dollar HeLa cell line.[66] But the Johns Hopkins researchers who first developed the HeLa line—although they took the tissue without the consent of Lacks or her family—had no commercial motives. They made the banked cell line readily available to other scientists without charge, in the old spirit to which John Sulston looked back nostalgically when he wrote about the way scientists used to see themselves as part of a global community with shared values.[67]

In contrast, biobank and database development is the DTC companies' principal strategy for commercial growth, many observers think.[68] Lori Andrews, a well-respected medical lawyer and professor, has even declared, of the direct-to-consumer genetic testing world, "Some companies are just a front end for biotech companies that use it for research."[69] If that's true, it would explain why firms are willing to sell the tests at a knockdown price or even to offer them for free, in the case of African Americans. Because black Americans are woefully underrepresented in genomic databases, their DNA could be a valuable resource for the company that can claim "brand edge" in it.

TruGenetics has given free DNA tests to the first ten thousand customers—regardless of race—willing to hand over their results for research. Likewise, 23andMe has offered a $99 test on the condition that the genetic analysis information remains in the firm's own biobank and that customers provide additional health and lifestyle data. By June 2011, 23andMe was able to announce that one hundred thousand customers had stored their genomic data with the firm, giving the company one of the world's largest genetic databases.

Why are these biobanks potentially valuable? Because the amount of variation in disease susceptibility contributed by each gene is typically so small (especially when epigenetics is taken into account), only mass databases can reveal statistically significant results. Conversely, whole-genome screening for large populations might eventually yield information that can be used to make a more accurate individual diagnosis.[70]

So the most promising aspects of genetic Me Medicine actually depend on We Medicine, in the form of collective databases. Particularly when the possible income from genetic patents is taken into account, these biobanks are a globally traded store of value: "biocapital."[71] Whether that capital should be shared collectively or held by individual firms is a controversy to which I'll return in the conclusion of this chapter and at greater length in chapter 7.

This value of biocapital in genetic databases is obvious from the explicit way in which retail genetics companies stake their claim on both the genomic data and the physical DNA banked with them. That might surprise most customers, who may well assume that they own their DNA even if they've stored it with the DTC companies. But while clients own the printout of their SNP analysis, they almost certainly have no rights in the biobank or in the DNA sample they stored there.

That principle was ratified in a 2002 court case called *Greenberg et al. v. Miami Children's Hospital Research Institute, Inc.* The parents of children with a fatal genetic condition called Canavan's disease had contributed tissue, information, and money to a database and biobank for the condition but were found to have no subsequent rights of control or ownership.[72] (This case is discussed at greater length in chapter 7.) Our law generally holds that once tissue has been removed or changes hands as a "gift," no further rights can be claimed by the donor. But just to eliminate any doubts, a "privacy statement" on the 23andMe website reads: "We may allow a commercial research organization access to our databases . . . so that . . . the organization can search without knowing the identities of the individuals involved, for the correlation between presence of a particular genetic variation and a particular health condition or trait. We may receive compensation from these research partners."

What's really at issue here is not your *genetic privacy* but the *private property* held by the testing firm in the database—which 23andMe unequivocally but legitimately calls "our" databases, you'll note. Precedents, including but not limited to the *Greenberg* case, have established an inequitable legal position: researchers, universities, and commercial companies can hold property in tissue and control its uses, but the individuals from whom the tissue was taken have no such rights. With such an imbalance of power in favor of commercial interests and against individuals, it's not paternalistic to think that those who already hold all the cards need to be subjected to some form of regulation; it's just realistic.

While the "no property" rule in excised human tissue originally dates back to the period before there was value in DNA biobanks or other body "products," the case that really set it in stone for the biotech age was *Moore v. Regents of the University of California* (1990). At the age of thirty-one, in 1976, John Moore had developed a rare cancer called hairy-cell leukemia. The condition required his spleen to be removed: that much was uncontested, since it had swollen to over twenty times its normal size. But Moore was also told by his surgeon, Dr. David Golde, that he needed to return frequently to donate samples of his hair, blood, sperm, and other tissue. Each time he was asked to sign a consent form, reading: "I (do, do not) voluntarily grant to the University of California [Golde's employer] all rights I, or my heirs, may have in any cell line of any other potential product which might be developed from the blood and/or bone marrow obtained from me."

Moore began by circling "do," even though he had doubts. "You don't want to rock the boat," he remarked in a later interview. "You think maybe this guy will cut you off, and you're going to die or something." But after Moore moved cities, Golde continued to insist that his patient must come down to Los Angeles for his "treatment," even though perfectly good hospital facilities were available in Moore's new home of Seattle. At this point, Moore's suspicions began to propagate. On his next visit, he circled "do not," resulting in a flurry of urgent phone calls from Golde's office instructing him to correct his "mistake."

It was then, in 1983, that Moore decided to take legal action, filing suit for "conversion" (unauthorized use of another's property). His lawyer, Jonathan Zackey, discovered that two years earlier Golde had already filed for a patent on the three-billion-dollar "Mo" cell line, which turned out to have unusually powerful and valuable immune cells. Although Moore was mainly incensed at the abuse of his trust, he was advised that his best chance lay in making a property claim.

But Moore failed in his legal action for conversion against Golde; his research associate Shirley Quan; the biotechnology firm Genetics Institute, Inc.; the drug company Sandoz Pharmaceuticals; and the Regents of the University of California. The final judgment from the California Supreme Court reiterated the common-law "no property in the body" principle. Judges siding with the majority expressed their fears that allowing tissue donors to have any rights in lines derived from their cells would inhibit scientific research and undermine human dignity, by creating a marketplace in body parts. In his dissent, however, Justice Broussard scoffed that tissue was *already* being valued in dollars and cents. Everyone stood to make a profit from Moore's tissue except Moore.

> Far from elevating these biological materials above the market, the majority's decision simply bars *plaintiff* [Moore], the source of the cells, from obtaining the benefit of the cells' value, but permits *defendants* [Golde et al.], who allegedly obtained the cells from plaintiff by improper means, to retain and exploit the full economic value of their ill-gotten gains free of the ordinary common law liability for conversion.[73]

This lack of say was extended beyond individual patients like Moore to entire groups of biobank tissue donors in *Greenberg* and another case, *Washington University v. Catalona* (2006). William Catalona, a respected urologist and surgeon who developed the prostate-specific antigen test for prostate cancer, had created a research biobank containing over 270,000 serum, blood, and DNA samples, along with 3,500 prostate tissue samples, much of which material came from his own patients. When Catalona decided to leave Washington University in St. Louis and take up a new post, he sent letters to all the patients he'd treated during his twenty-five years there. Six thousand men agreed to

Catalona's request that their samples should move with him to his new job at Northwestern University so that he could carry on with his research.

But Washington University went to court, claiming that the men's samples belonged neither to the donors nor to Catalona himself but to the university, as Catalona's employer. Both the district and the appeal courts agreed with the university, dismissing the statements made in court by some of the men about why they wanted their tissue to go with Catalona: they trusted his work and hoped he could find a cure. The Supreme Court declined to hear the case, so that decision is final—and it has important implications for who owns genomic data in DTC genetic testing biobanks. Not the individual donors, is the likely answer.

One patient, Richard Ward, had said, "Washington University was where Dr. Catalona was, so that's where I was [for my operation]." Another man, James Ellis, declared: "I have six grandsons and the one thing I want to do is what I can to make certain they don't go through what I've gone through, and my family's gone through, for the last fourteen years. And I [can't] think of anybody that I would have more faith in to do the kind of research that might help my grandsons on my samples, my tissues, my body parts, than Dr. Catalona."[74] But what James Ellis didn't realize—as most people probably don't—is that the no-property rule meant that the biobanked materials were no longer *his* samples, *his* tissues, or *his* body parts. The courts ruled that because the consent form that the patients had signed was on stationery headed "Washington University," the men should have realized that they were permanently transferring ownership of their tissue to the university, not to Catalona. On the issue of "whether individuals who make an informed decision to contribute their biological materials voluntarily to a particular research institution for the purpose of medical research retain an ownership interest allowing the individuals to direct or authorize the transfer of such materials to a third party," the judgment declared, "the answer is 'no.'"[75]

Catalona illustrates how few powers tissue donors have over "downstream" uses made of their cells, especially when commercial interests are at stake. In this case, a substantial coalition of academic institutions and medical researchers filed *amicus curiae* (advisory "friend of the court") briefs on behalf of Washington University, asserting that their business interests would be threatened if researchers such as Catalona could take valuable databases with them or if donors were given a say. Those bodies included the American Cancer Society, the Mayo Clinic, the American Council on Education, the Association of American Medical Colleges, and the Association of American Universities, along with many individual medical colleges, including Stanford, Cornell, and Johns Hopkins. Catalona had already been rapped over the knuckles for altruistically proposing to share two thousand samples with researchers from another university—the sort of thing we naively imagine that disinterested medical researchers do all the time, in their search for scientific truth and human betterment. "Just from a cost recovery scenario," the university's business manager had scolded in a memo to the vice chancellor for research, "this should be worth nearly $100,000."[76] In one of the very few cases to hold that the common law *does* confer a property right in excised tissue, *Yearworth v. North Bristol NHS Trust*, there was no commercial value in the tissue involved, namely, stored samples of several men's sperm taken before their operations for cancer and subsequently ruined through negligent storage by the hospital. Although some commentators think that this 2009 English judgment "signals a sea change in judicial attitudes toward patients' rights,"[77] the circumstances only pitted a handful of patients against a single not-for-profit hospital. Additionally, this was a case in which there was a clear intention by the patients *not* to donate the tissue unconditionally to someone else but to retain it for their own personal future use. That scenario didn't apply in *Catalona* and probably wouldn't be true of any case involving a DTC genetic testing company.

And, of course, the *Yearworth* judgment isn't binding in the United States. While U.S. judges could draw on the reasoning of the British judges in *Yearworth* if they so chose, they aren't obliged to hold by its verdict. However, the reasoning is certainly relevant on the western shores of the Pond. The English court said: "*In this jurisdiction* developments in medical science now require a re-analysis of the common law's treatment of an approach to the issue of ownership of parts or products of a living human body."[78] Those developments in medical science certainly aren't confined to England. Indeed, commodification of genes is further advanced in the United States than in the United Kingdom, particularly where genetic patenting is concerned. Many U.S. genetic patents aren't valid in Europe, and the NHS has taken a conscious decision to ignore the *BRCA1* and *BRCA2* patents because they're felt to impede national patient care.

In the United States, the issue of genetic patents has recently been contested in *Association of Molecular Pathology v. Myriad Genetics*. A coalition of concerned professional organizations, physicians, patients, and the American Civil Liberties Union successfully obtained a district court judgment against the restrictive *BRCA1* and *BRCA2* patents in early 2010.[79] That judgment was overturned on appeal in July 2011, but only in a split 2–1 decision, so both parties petitioned the Supreme Court to settle the issue. In March 2012, however, the court granted *certiorari* (review) only long enough to instruct the Court of Appeals for the Federal Circuit to reconsider its decision in light of another case that had just been decided, *Mayo v. Prometheus*. But on August 16, 2012, the appeals court upheld its original decision that the *BRCA1* and *BRCA2* genes as patented by Myriad were not a natural composition of matter but rather a manmade product. In November, however, the Supreme Court agreed to review the case in 2013.

While cases determine policy in common-law systems, so, of course, do statutes. That's why the Vermont and Massachusetts bills could be highly significant in establishing who owns the genome and genetic information. We've already seen that the two bills both propose to make genetic information "the exclusive property of the individual from whom the information is obtained." That covers the data, but what about the banked DNA itself? The proposed Vermont legislation goes on to confer upon genetic material the status of "real property subject to one's individual control and dominion in accordance with generally held precepts of property law in Vermont."[80] But the "generally held precepts of property law in Vermont," as in the rest of the English-speaking world, traditionally have *denied* that tissue taken from the body is like real estate or other forms of personal property.

So that provision of the Vermont bill really would be a major change from the common-law position—as would the way the Massachusetts bill contemplates genetic information being made into heritable property. Under its terms, people can bequeath to their surviving spouse or any other legatee the right to use their genetic information. If heirs could withdraw the data at will, that would radically undermine the exclusive rights that retail genetics companies now hold in their databanks. The resulting uncertainty would certainly be a disincentive for other firms to buy into the DTC companies' biobanks, if they had no way of knowing exactly which files and what information they were purchasing.

Even more fundamentally, the Massachusetts bill recognizes the inherent monetary value of genetic information.[81] Section 1(b) stipulates that before entering into a contract to share genetic material or genetic information—presumably including a contract with a DTC company—people must be notified orally and in writing that "their donation is a commodity and is of some material value." (A similar provision would apply in Vermont.) Furthermore, if the biobank intends to commercialize the genetic information, the individual donor in both states must be made aware and compensated at a fair market value.[82]

This provision threatens the retail genetics companies' business plans—and to be fair, it's hard to see how the companies could know in advance what a reasonable market value would be. There needs to be some sort of retrospective way of determining what amounts donors are owed after pharmaceutical firms or other commercial users of biobanks have drawn up their contracts with the DTC companies. Before that happens, however, you can be pretty sure that the retail genetics firms' attorneys will be dusting off their copies of *Moore*. Of course genomic DNA "is of some material value"; so was Moore's tissue—to the tune of $3 billion. But that wasn't enough to turn it into his property, as the California Supreme Court held.

If the Vermont and Massachusetts bills go through, they will almost certainly be challenged in court, raising dilemmas about precedence of statute over case law and issues about contending state jurisdictions. Whether they will raise a constitutional issue and make it into federal courts is another interesting question. Given how fundamental the no-property rule is in the common law and how influential are the business interests involved, it seems likely that the issue would be pursued, but it's unclear what constitutional rubric would cover it.

While DTC genetic tests are marketed with a message touting individual responsibility—as the poster child for Me Medicine—databases and biobanks derived from many people's contributions could be seen as rightfully as a We resource, a form of commons. Already the retail genetics companies use "We" as a marketing message, along with "Me": 23andMe, for example, promotes "sharing and community" by setting up a social network of its users—called, notably, 23andWe. "Our features also give you the ability to share and compare yourself for family, friends and people around the world," the website promises.[83] But this supposed commitment to community anything more than a way of increasing market share and widening the customer base? It does nothing to change the ownership of the data, which remains private to 23andMe and the business partners whom the firm permits to mine its databases.

More genuinely communitarian associations than any commercial firm—groups of people with common genetic risk profiles for particular diseases—have started to take advantage of the "We"-ness of the Internet. One such group is FORCE (Facing Our Risk of Cancer Empowerment), a forum and advocacy network for women with a family history of breast and ovarian cancer relating to the *BRCA1* or *BRCA2* genes. Another genetic interest group, for parents of children with the genetic condition PXE (pseudoxanthoma elasticum), has pioneered a joint ownership model of patent rights with a commercial firm, plowing profits back into further PXE research. In both these cases, however, only specific medical conditions are involved, and thus the groups are limited in size and influence.

David Winickoff, a Berkeley political scientist who has devised a new model of charitable trusts for biobanks that has been taken up by many commentators,[84] thinks that even heterogeneous groups such as DTC customers could have some communal property rights, if there's sufficient political and commercial will. For example, they could have something like representatives on a shareholder's association, even if they don't actually hold equity shares. That would certainly give them a lot more say than Catalona's patients or 23andWe's social network, but will it happen?

Given the view from leading medical bodies that DTC genetic testing has few benefits for the donor no matter how low the price, customers could and should be protected by consumer legislation. Alternatively, given that the common law has traditionally been loath to view individuals as owning their bodies and that regulatory bodies have been somewhat slow off the mark, it might be more effective and legally coherent to limit retail genetics companies' untrammeled downstream rights in biobanks. The Vermont and Massachusetts bills may contain contradictions, but they do demonstrate that there is some political momentum away from simply letting the market decide and toward recognizing that opting for retail genetic testing isn't just an individualistic consumer choice.

THREAT, NARCISSISM, CORPORATE INTERESTS, AND CHOICE RECONSIDERED

Let's now revisit those four possible reasons why Me Medicine is edging out We Medicine, applying them to personalized genetic testing.

The first hypothesis introduced in the previous chapter concerned a society-wide sense of *threat and contamination*. Toxic threat is actually much more intimately linked to the history of the Human Genome Project than most people realize, according to the documents uncovered by Helen Wallace and others. They allege that the tobacco companies intended to camouflage the threat by spreading the word that the risk was confined to genetically predisposed people—those whose genes made them susceptible both to addiction to nicotine and to the carcinogens in tobacco smoke.

Wallace goes so far as to say that the origins of the Human Genome Project suggest that personalized medicine began life as a public relations message invented by the tobacco industry and was promoted to undermine public health strategies such as smoking prevention.[85] Although this may sound like a conspiracy theory, it's not a theory at all but rather the outcome of careful and systematic gathering of empirical evidence. However, the rationale of the HGP was also genuinely entwined with concern about genetic damage to public health from toxins such as tobacco smoke and radiation. When the burden was shifted to susceptible individuals, that communal concern was lost.

In the modern context of personalized genetic testing, the sense of threat is sometimes a factor. We've seen that leading proponents of personalized genetic testing play on the sense that everyone is at risk, as George Church put it. Avoiding the threat to the unborn child from undiagnosed genetic disorders in the prospective parents is implicit in the marketing message of Counsyl's genetic testing service. Although that might seem like a responsible thing to do, the British consultant geneticist Frances Flinter remarks of this development, "It plays unnecessarily on people's fears."[86]

On the other hand, optimism rather than fear is the brand for DTC companies selling genetic matchmaking services—such as Scientific Match, which promises to put you in touch with partners whose genetic makeup will supposedly enable you a better sex life and a high natural immune response in any children you may have together. Boosterism is also rampant in the claims from companies offering to test your children for the *ACTN3* gene allele involved in athletic performance.[87] More generally, the retail genetics companies' websites are imbued with a sense of excitement about scientific progress, of which you too can be part just by sending in a spit sample and paying your fee.

So although some customers of DTC testing services, such as David Ewing Duncan, may be motivated by a sense of environmental contamination, on the whole, promise seems far more central than threat in explaining the rise of retail genetics. But is that promise inflated? I'll examine that question at greater length in chapter 3, which discusses pharmacogenomics, probably the area that comes closest to fulfilling the promises of personalized medicine. For now, it's worth noting that while individualized genetics more generally—not just DTC testing—can claim advances in identifying disease pathways, diagnostics for a limited number of single-gene disorders, and some early success in tailoring treatment regimes, it still falls short of widespread clinical applications.

Continuous discoveries of new surprises about the genome call into question the claim that personalized medicine is almost here, or that individualized drug therapy will soon be a reality. In fact, it probably never will be, or at least not by DNA testing alone, because most genotype-phenotype associated studies are hampered by limited size and therefore decrease in statistical power.[88]

The real threat, in my view, is that despite its imperfect evidence base, personalized genetic medicine will edge out the more pressing needs of public health: that Me will shove We aside. President Obama's proposed budget for 2012 committed the federal government to a 2 percent overall cut for the Department of Health and Human Services—the first in the department's history, despite our graying population—but a 2.4 percent increase for the National Institutes of Health. Francis Collins of the NIH planned to use the funds to focus on "leveraging new genomics technologies in disease and health research and translational science, and in pursuing goals in personalized medicine."[89] But Obama's budget also proposed a staggering 90 percent *decrease* in the budget of the Centers for Disease Control's Office of Public Health Genomics. As the medical lawyer Jonathan Kahn explains:

> The OPHG conducts some very valuable population-based analyses of the role of genomics in improving the public's health. For example, it recently funded the Michigan Department of Community Health to increase the number of health plans that have policies consistent with U.S. Preventive Services Task Force recommendations for genetic risk assessment for hereditary breast and ovarian cancer. Public health genomics, however, is not a money maker. The sort of research supported by the OPHG does not lead to new products that can be developed and marketed by large pharmaceutical corporations.[90]

Yet it's the Centers for Disease Control that deal with such genuine and massive threats to public health as swine flu and other pandemics. (In chapter 6, I'll examine those threats, and the virulent reaction against vaccination, at greater length.)

Some threats, it seems, are more equal than others. If threat does go anywhere toward explaining the rise of Me Medicine, it's because threat has already become individualized in many people's minds: threats to *my* baby or *my* own health, for example, rather than a viral epidemic that could threaten all of *us*. But what explains that individualization in the first place? What about the second possible explanation, narcissism and "bowling alone"?

Narcissism works best as an explanation for "recreational genetics" but less well for the one-third of customers who think that they're buying a diagnosis. Those individuals aren't necessarily in the fell grip of a narcissism epidemic; they're just misled about the effectiveness of SNP testing as compared with more traditional measures, such as systematic family history taking.[91] This theme will recur in later chapters: other forms of personalized medicine, particularly private umbilical cord banking, are also medically less effective than the more community-minded alternative (in that case, public cord blood).

"Me" often markets itself as clinically superior to "We": that's an essential premise of personalized medicine. The evidence base doesn't always bear that claim out, but people may not be aware of the evidence. That doesn't make them narcissistic; it just means that they're not as well informed as they need to be. Part of this book's mission is to examine that evidence base so that people can make themselves better informed.

Nor is it narcissistic to be confused about the difference between predicting with certainty whether *I as an individual* will develop a particular disease and estimating the probabilistic susceptibility for a *population*. Direct-to-consumer genetic tests identify our disease risks through probabilistic assessments, based on our genetic similarities to others with common diseases that have a known incidence, as derived through genome-wide association studies. So actually they rely on "We"-ness even as they proclaim their "Me"-ness.

> Scans are personal in the usual sense that, like much of what doctors tell us, they are about ourselves. . . . This is a geneticized medicine which predicts our future health and disease. But it attempts to do this by pointing not to the individuality of our genetic natures, but to our genetic similarities with others. It identifies our disease risks through our genetic similarity with others with a known experience of common disease.[92]

The related "bowling alone" hypothesis works a bit better than narcissism, if you consider the way in which 23andWe or the recreational ancestry services claim to be forging a new kind of connectedness.[93] Like the recent rise of alternative kinds of civic togetherness that Putnam identified after the 2008 election, those forms of genetic testing could be seen as a new form of social network. Initiatives such as "Roots Into the Future," a free retail genetic testing offered by 23andMe to ten thousand African Americans, may reach minority ethnic groups who never fitted that well into the benign picture of 1950s togetherness, as Putnam himself admits.

The legal scholar Dorothy Roberts believes that most of her fellow African Americans aren't interested in finding out about their ancestry in America and often can't do so even if they try, because no one kept birth and death records for slaves.[94] But Alex Haley's bestseller *Roots* tapped into their deeper desire to trace their African ancestry. "Roots Into the Future" plays on both Black Pride and the appeal of cutting-edge technology, drawing in potential customers for more complete and more expensive versions of the testing service. On the other hand, it's also entirely plausible to interpret "Roots Into the Future" as a way of extending market share for the companies involved while claiming to promote togetherness. And the resulting biobank of comparatively rare African American genotypes is valuable biocapital.

Corporate interests and neoliberalism, the third hypothesis, strike me as much more powerful explanations of retail genetics than narcissism. The Canadian scholar Roxanne Mykitiuk believes there's a significant affinity between the "new genetics" and the central neoliberal policy of privatization—not just in the sense of outsourcing previously government-run services to the private sector but also of privileging the private above the communal.[95] The entire premise behind genetic testing—not just the economics—can be seen as neoliberal: "The diversion of attention from social to molecular causes and solutions reinforces privatization, the hallmark of the neo-liberal state that pervades every aspect of public policy."[96]

But let's be a little more precise: taking part in that area of Me Medicine known as retail genetics doesn't necessarily mean abandoning your commitment to your regular healthcare provider, which may be partly or even wholly publicly funded (as with Medicare or the NHS). The problem isn't so much the private sector replacing the public as the public being overwhelmed by additional demands created by private DTC companies without being given any additional resources—even while suffering major cutbacks under austerity programs. In other words, the public will have to prop up the private again, which many commentators believe to be the reality, though not the rhetoric, of neoliberalism.[97] One reason why doctors are concerned about retail genetics, as we've seen from the professional guidelines, is the potential overload on the medical profession, health insurers, and a national health service like the United Kingdom's.

Neoliberalism is sometimes defined by its critics as socialism for the corporations and the free market for individuals. That's simplistic but catchy—

and quite appropriate to the way in which private capital has relied not only on a permissive U.S. regulatory regime in patenting but also active government backing in developing the blue-sky science that can then be "translated" into profit-making services such as retail genetics. True, the private sector also provides venture capital, for example, through Google's links with 23andMe. But deCODEme relied on the initial public resource of the national Icelandic database, which the government made available to the firm on an opt-out basis: it was assumed that Icelanders had allowed the use of their data unless they explicitly withdrew.

Developments in retail genetics mirror a more general trend in biotechnology and biomedicine, according to Mariana Mazzucato, a professor of science and technology policy at the University of Sussex and the author of *Risks and Rewards: Understanding the Innovation-Inequality Relationship*:

> Where would GSK and Pfizer be without the $600 billion the US National Institutes of Health has put into research that led to 75% of the most innovative new drugs in the last decade? The state's role . . . was not just about correcting "market failures." What the state did was to take on the greatest risk, before the private sector dared to enter—acting as an "entrepreneurial" state. In biotech, venture capital entered 15 years after the state invested in the biotech knowledge base. . . . In biotech, venture capital has entered late and made a killing from an industry it did not create.[98]

Given that the state bore the financial risks, you might think it should have some say in regulating biotechnology industries, but that's not how it works. The antiregulatory agenda adopted by the UK's Human Genetics Commission and Nuffield Council is typical of much policy response in advocating no more than voluntary self-regulation by the retail genetics industry. As I've argued already, and as the British political scientist Stuart Hogarth agrees, when surveys show that people want regulation, "It's a paternalistic neo-liberal agenda to say they're somehow misguided."[99] Oddly and ironically enough, he says, industry actually claims to want a code of practice. Contravening the wishes of both the public and the industry in the name of market freedom bespeaks an ideological platform such as neoliberalism rather than a practical strategy.

Yet as we've seen, roughly half of U.S. states do regulate retail genetics, which the neoliberal hypothesis wouldn't predict. Along with regulation by the FDA and individual states, corporate ownership rights in human tissue and genetic patents are being challenged by individual court cases such as the *Myriad Genetics* lawsuit. Neoliberal ideology doesn't rule unchallenged in the political realm. It would be a simplistic mistake to attribute the success of Me Medicine to that and that alone—although it's also a salutary correction to the way in which it's so rarely mentioned in the usual treatment of buying genetic tests as a purely individual choice.

In terms of the fourth hypothesis, individual *choice*, personal autonomy, and individual self-discovery are certainly foregrounded by many writers of personal genomic odysseys.[100] Martin Richards, however, accuses retail genetics of actually *undermining* autonomy. He argues that the DTC companies claim to be able to tell us more about our innermost selves than we can know through our own cognition. Likewise, wrong information, in the form of a false positive or false negative, could easily undermine rather than enhance rational choices and a sense of being master (or mistress) of your own health. The sense of being a patient from cradle to grave also weakens our sense of personal agency rather than enhancing it.

So choice and autonomy are not actually being served, despite the retail genetics companies' rhetoric claiming "genetics just got personal." But we need to go further, moving from the descriptive (*is* autonomy actually the key value in personalized medicine?) to the normative (*should* autonomy be the key value that we want to see medicine serve?). As I suggested in chapter 1, there's been a reaction against that assumption in recent writings on medical ethics. Skepticism about whether autonomy is paramount hasn't really spread beyond academic circles, however, and even there, it's probably a minority view. As other commentators have also argued,[101] autonomy and choice are still largely regarded as knockdown arguments in general public debate.

I've argued that in fact, these quintessential Me Medicine values don't fit all that well even where they're very strongly touted, as they are in retail genetics, and that they shed more heat than light—no matter how central they are to brand strategy for the DTC firms. They form a poor fit with the reality of who has property rights in and autonomous control over the stored tissue sample. Maybe "your genetic information should be controlled by you," or maybe not: that's a normative question, with the "should" giving it away. Descriptively, however, that's not how it really works in retail genetics. The language of individual ownership of your health largely masks the collectivization and privatization by firms of your genetic data.

In the final chapter, I'll develop a more complex analysis of how communal values, such as the notion of a genetic commons in which we all share, might inform medicine more satisfactorily than autonomy and choice. Before doing so, in the penultimate chapter I'll look at case examples indicating that our devotion to choice and autonomy, in the instance of vaccination refusal, potentially compromises both our ability to achieve those We Medicine values and our health itself. But both those chapters are some way down the line. Let's move on now to the second practical example of Me Medicine: pharmacogenetics and pharmacogenomics.

3

PHARMACOGENETICS: ONE PATIENT, ONE DRUG?

In 2004, at the age of seventy-one, the author Susan Sontag died of cancer—the disease that she had identified as the emblem of modernity in her book *Illness as Metaphor*. Poignantly and presciently, Sontag had called cancer "obscene—in the original meaning of that word: ill-omened, abominable, repugnant to the senses."[1]

Cancer is modern both in a symbolic sense—"a disease of overproduction, of fulminant growth"[2]—and in the factual medical sense. Now that the epidemics of the early industrial age—cholera, typhus, smallpox, diphtheria, typhoid—have been decimated by effective We Medicine programs such as sanitation and vaccination, we die later, but increasingly we die of a disease that strikes us as isolated individuals, one by one. (True, some cancers, such as Burkitt's lymphoma, have a viral link and occur in epidemiological swaths, but they're the exception.)

Sontag's death from cancer is common knowledge. What is less well known is that she died neither of her original, naturally occurring uterine cancer nor of her later encounter with breast cancer. Instead, Sontag fell prey to an iatrogenic cancer—bone marrow myelodysplasia leading to leukemia, caused by high-dose chemotherapy to treat her earlier malignancies.[3]

Many such cases resulted from the then-orthodox "megadose" regime, castigated by the oncologist Siddhartha Mukherjee in his Pulitzer Prize–

winning "biography" of cancer, *The Emperor of All Maladies*. Mutations induced by chemicals in chemotherapy or x-rays in radiation therapy can cause cancer by activating proto-oncogenes dormant within the patient's body. Another form of iatrogenic harm, Mukherjee believes, was the previous but now discredited orthodoxy of radical mastectomy for breast cancer. Yet despite these scorched-earth campaigns, between 1962 and 1985 cancer-related deaths in the United States actually *rose* by 8.7 percent.[4]

Such aggressive but ultimately pointless interventions might well seem to epitomize the worst of the one-size-fits-all approach to cancer care. This has led many to argue that it's not only more humane but also medically more effective to individualize treatment, so that patients can be spared overtreatment that will do them more harm than good. It might be possible, for example, to identify patients who are genetically programmed to respond more quickly to chemotherapy and to give them lighter dosages. Then "kill or cure" wouldn't translate, as it did in Sontag's case, to "cure but kill."

This is the powerful premise of pharmacogenetics,[5] the relation of heritable variation to individual differences in drug response.[6] Pharmacogenetics aims to tailor pharmaceutical regimes in cancer treatment and other branches of medicine to the patient's individual genome. The goal is to minimize the adverse drug reactions estimated to cause ten thousand deaths a year in the United Kingdom alone, with even more patients suffering serious side effects.[7] Pharmacogenetics isn't confined to oncology, but there the goal is also to adjust treatment to the sequenced genome of the cancer, which differs from the patient's normal cells.

This double approach is crucial because cancer is so heterogeneous, even in patients with the same diagnosis. After sequencing the entire genomes of fifty patients' breast cancers, researchers found that only 10 percent of the tumors had more than three mutations in common.[8] Breast cancer is actually ten different diseases when analyzed genetically, according to a recent study of two thousand tumors that mapped types of genetic mutations and SNPs.[9] If "the cancer cell is a desperate individualist"[10]—dissimilar to the cells that produce cancer in other individuals diagnosed with the same disease and uncivilly disobedient to the body's rule—shouldn't it be treated in an individual fashion?

Powered by a genuine threat from a wily and vicious disease, personalized cancer medicine in the form of pharmacogenetics could well appeal even to those who view direct-to-consumer genetic testing as a trivial and inaccurate bastardization of genetic expertise. The standoff between Me Medicine and We Medicine might be subject to the arbitration of pharmacogenetics. Some commentators, for example, believe that investment in pharmacogenetics is the most effective way to deliver the public health benefits originally promised by the Human Genome Project.[11]

On the face of it, there's nothing intrinsically Me-ish about genomic medicine. A major public health study—the five-year "Human Heredity and Health in Africa" (H3) study, jointly funded by the NIH and the Wellcome Trust—aims to apply genome scanning and sequencing techniques to major communicable diseases such as HIV/AIDS, tuberculosis, and malaria as well as to noncommunicable conditions such as cancer, stroke, heart disease, and diabetes. The hope is that the project will finally bring some of the benefits of advanced genetics research to the world's poorest continent. Additionally, because all human populations derive from Africa, and because Africa contains greater genetic diversity than any other continent, information derived from the study of African populations will enhance our understanding of diseases around the globe. More pessimistic analysts, however, believe that such high-profile initiatives will be few and far between and that there is a deeper inherent tension between the collectivist principles behind public health and the individualistic slant of personalized medicine.[12]

Initially, there seems less reason to be skeptical about pharmacogenetics than about retail genetic testing. The research evidence base is far stronger and growing all the time. In February 2012, for example, vemurafenib (Zelboraf), a gene-specific drug for aggressive melanoma (skin cancer), showed promising findings in a clinical trial reported in the *New England Journal of Medicine*.[13] The drug, which targets a genetic mutation in the tumors of approximately half of all melanoma patients, didn't cure the cancer but did extend the lifespan of patients by sixteen months, compared with nine months on conventional treatment. Only about half of patients with the genetically "right" type of tumor responded to the drug, however, and in turn the mutation only occurs in half of all melanoma patients, so that only one patient in four could benefit. (As you can imagine, a BBC television program about the new drug focused on that one in patient in four and didn't say anything about the other three.) Still, for those patients this pharmacogenomic personalized therapy could make an important difference in giving them a few more months of life.

However, while there may be less reason to be skeptical about the promises of pharmacogenetics than about those of retail genetic testing, that doesn't mean there is no reason. We need to bear in mind the possibility that pharmacogenetics may be just the latest "magic bullet" in the history of oncology.

Mukherjee calls his own profession "a branch of medicine particularly suffused with hope (and thus particularly prone to unsubstantiated claims of success)."[14]

A month after the melanoma study, the *NEJM* reported much less encouraging results for pharmacogenetics in a study from the Royal Marsden Hospital in London, the well-known center of excellence for cancer care and research. A genome-wide analysis of biopsies done on four kidney cancer patients showed that that a single tumor can have many different genetic mutations at various locations. Two-thirds of the genetic faults identified weren't repeated in the same tumor, let alone in any other metastasized tumors in the body.

That's quite discouraging, because if a pharmacogenetic drug targets one mutation in the tumor, it won't necessarily work on other mutations. And that's assuming further mutations don't occur, though they're likely to if the cancer figures out ways around the drug. As the *NEJM* article put it in quite chilling fashion, "Intratumor heterogeneity . . . may foster tumor adaptation and therapeutic failure through Darwinian selection": survival of the fittest cancer genomes within the tumor, threatening the patient's own survival.[15]

It's this chapter's job to come up with a balanced judgment about pharmacogenetics, taking into account not only the medical evidence but also issues about justice, patenting, and drug rationing—including dilemmas about racial equality that arose with one particularly controversial "personalized" drug, BiDil. Throughout, I'll be testing the promise of pharmacogenetics against its logical (but commercially and scientifically untenable) extreme: one person, one drug—the ultimate in Me Medicine.

EVALUATING PHARMACOGENETICS: THE POSITIVE EVIDENCE

In 2011, a team of researchers at Washington University in St. Louis announced their success in applying whole-genome sequencing from skin and bone marrow DNA to the care of a female cancer patient. Despite the lack of any significant family history of cancer, the patient had early-onset breast and ovarian tumors. She also had acute myeloid leukemia—triggered by her treatment for the other cancers, like Susan Sontag's. But what sets this case apart from Sontag's is that this woman's therapy could be adjusted to her genetic profile. As a result of unsuspected genetic defects revealed by whole-genome sequencing, the patient's treatment plan was changed from bone-marrow transplantation to targeted chemotherapy. She is now in remission.[16]

A year earlier, on the basis of a partial genetic sequence, a Wisconsin boy was given a risky bone-marrow transplant that might not otherwise have been contemplated. He, too, is now in remission.[17] While the decisions about how to treat turned out to be opposite in the two cases—don't proceed with a transplant in the St. Louis case, do proceed in the Wisconsin one—both exemplify the pharmacogenetic method: using whole or partial genomic sequences to tailor clinical care to the individual patient.

Childhood leukemia treatment has been turned around by a blood test to identify tolerance for the "miracle drug" 6-mercaptopurine, which cures over 80 percent of affected children but also kills some whom it might have cured had it been given in smaller doses. A genetic test is now available to separate out those children who are most at risk of dying because of the drug, those of normal susceptibility, and those who will probably not respond at any dosage level.[18]

More recently, researchers discovered a gene mutation that can increase the risk of developing blood clots for breast cancer patients being treated with tamoxifen, allowing affected women to decide whether to choose another therapy.[19] Likewise, women considering tamoxifen treatment can now be genetically classified as probable poor, intermediate, or rapid metabolizers, and those results can be incorporated into clinical decision making.[20]

In all these cases, it's the *patient*'s genome that's been sequenced. Unraveling the *cancer*'s genome—the second prong of pharmacogenetic strategy in oncology—is best exemplified by the anticancer drug Herceptin (trastuzumab).

In the early 1980s, the geneticist Robert Weinberg and his postdoctoral researcher, Lakshmi Charon Padhy, isolated an oncogene from a rat tumor, which they linked to the overproduction of cancer cells. Subsequently, another team of researchers, led by Axel Ullrich at the biotechnology company Genentech and the UCLA oncologist Dennis Slamon, noted the rat gene's resemblance to another growth-modulating gene. This gene, later christened *HER-2* (Human Epidermal Growth Factor Receptor-2), causes "overexpression" or "amplification" of breast cancer cells—in other words, it makes them grow fast and furiously.

Ullrich and Slamon discovered in 1986 that breast cancers could be divided into those that were negative for the *HER-2* gene and those more virulent tumors that were positive. In cases of *HER-2*-positive breast cancer, tumor cells depend for their growth on this gene, which in turn spurs their growth by making multiple copies of itself. By blocking the gene from expressing proteins in *HER-2*-positive patients, it might be possible to tackle those cancers. Although Genentech management pooh-poohed the profit-making potential of a drug targeting the gene, Slamon's persistence finally resulted in a successful clinical trial and the development of the drug Herceptin. One of the original trial subjects, Barbara Bradfield, has now been stable on the drug for more than twenty years, even though her original tumor had spawned sixteen new masses in her lung by the time the research began.[21]

Tumor sequencing has developed apace since the Herceptin trial. In 2010, the United Kingdom's National Health Service announced plans to personalize the treatment of up to six thousand patients annually by adjusting their therapy to the genetic mutations of their tumors. Announcing the initiative, the chief executive of Cancer Research UK, Dr. Harpal Jumar, remarked: "We believe that cancer medicine has reached a point where increasingly the genetic characteristics of individuals' tumors will and should dictate what treatments they receive. We now have enough genetic markers and drugs for this to make a real difference. It's patently obvious that this is going to be the way of the future."[22]

The sort of evidence to which Jumar is referring may well come from studies such as the current five-year international Cancer Genome Project, which will compare the responses of 350 tumor samples to eighteen drugs used to fight cancer, or the Norwegian Cancer Genomics Consortium, which is using "next-generation" DNA sequencers to trawl for mutations in common tumors.[23] The Stratified Medicine Programme, launched by Cancer Research UK in November 2011, is setting up a routine genetic testing service for cancer patients, incorporating the data into both clinical care and a central research database. Initially it is concentrating on breast, bowel, lung, prostate, ovarian, and skin cancers.[24]

Already, genetic tests of breast cancers can be used to predict accurately which patients' tumors are not likely to metastasize, potentially sparing them aggressive chemotherapy after surgery.[25] For non-small-cell lung cancer, a test called SNaPshot (with its name derived from "SNP") has been used to identify two common mutations in 51 percent of samples and to enroll a subgroup of patients into a clinical trial for a gene-targeted drug.[26]

Outside the cancer ward, there have been other signal successes for pharmacogenetics.[27] On the research front, the Royal Brompton Hospital in London announced in 2010 that it would begin sequencing the exomes (DNA-coding portions of the genome) of ten thousand patients with heart disease. By reducing the cost of sequencing compared to the whole genome, concentrating on the exome allows the researchers to cover a large number of patients over a ten-year period. The goal is to determine genetic links to subtypes of cardiovascular disease and abnormal variations, such as enlarged or unusually heavy hearts. Likewise, the U.S. National Genome Research Institute has moved into whole-exome sequencing of arteriosclerosis in a trial involving 1,500 patients.[28] Another research project is going on at the Mayo Clinic, which is using whole-genome sequencing in testing a group of patients for eighty-three genes that determine how the body metabolizes a wider range of drugs, not just those used against cancer.[29]

In pediatrics, a seven-year-old boy with a severe gastrointestinal disease similar to Crohn's syndrome was diagnosed by sequencing his exome. He was found to have an X-linked condition leading to life-threatening cell death. The boy had a bone-marrow transplant following this molecular diagnosis; seven weeks later, he was able to eat and drink normally, with no recurrence so far of his gastrointestinal disease. As the authors say, this case "demonstrates the power of exome sequencing to render a molecular diagnosis in an individual patient in the setting of a novel disease, after all standard diagnoses were exhausted, and illustrates how this technology can be used in a clinical setting."[30]

Whole-genome sequencing was used to refine a diagnosis of dopamine-responsive dystonia, a rare neurological condition similar to Parkinson's disease, in two fourteen-year-old fraternal twins, Noah and Alexis Berry.[31] Researchers and physicians at the Baylor Genome Sequencing Center and the Texas Children's Hospital traced the twins' illness to a gene that plays a crucial role in serotonin production. They added a serotonin-inducing supplement to the twins' diet, eliminating many of their symptoms and even enabling Alexis to start running competitively. Pediatricians at the Clinic for Sick Children in Pennsylvania, aided by molecular geneticists, are also integrating genomic techniques into pediatric care of the Amish and Mennonite communities, which have particularly high rates of certain genetic disorders.[32]

Treatment for the liver disease hepatitis C was also successfully personalized in June 2011. That means that the worst side effects can be avoided for some patients, those whose variation in an SNP located near the gene *IL28B* makes them more responsive to a lower dosage.[33] The variant seems to change the relevant protein's interaction with the interferon-alpha used in treatment. This study was the first to use genome-wide association analysis (rather than whole-genome sequencing) to translate results from bench to bedside. How-ever, a potentially controversial finding was the ethnic difference in response rates to the treatment, which were lowest for African Americans and highest for patients of Asian extraction. (Later, I'll return to the torrid question of ra-cial prescribing, which has been criticized as biologically implausible and po-litically explosive.)[34]

Another use of pharmacogenetics outside the realm of oncology lies in fertility treatment. There's been considerable concern about the long-term risks of ovarian stimulation drugs given to women undergoing fertility ther-apy or selling their eggs for IVF.[35] One study of 15,000 women who had taken drugs for ovarian stimulation found a statistically significant increase in uter-ine cancer and a borderline significant risk of malignant melanoma, non-Hodgkin's lymphoma, and breast cancer.[36] Fatalities or near-fatalities have re-sulted from overstimulation, including a shocking case in which a young woman was given such high dosages that she produced seventy eggs and nearly died.[37] When you consider that the human female is programmed by nature to produce one or at most two eggs per cycle, this looks suspiciously like a serious form of iatrogenic harm. If clinicians could identify genetic-linked differences in response to ovarian stimulation, they could use lower dosages for highly re-sponsive women and minimize the real risks from overstimulation.

For several years, it's been known that one particular gene, *FSHR*, is linked to response to ovarian stimulation.[38] Further progress in this direction was announced in 2010 by a Yale University Medical School team, which found a genetic mutation in an abnormal receptor on the surface of cells in the egg follicle, affecting response to follicle-stimulating hormone.[39] If translated into clinical practice, these findings could lessen the risk of potentially fatal ovar-ian hyperstimulation syndrome.

There could be considerable impact on the developing world of the discov-ery that a "Goldilocks" gene affecting response to tuberculosis can suggest the best treatment. "Depending on what versions of the *LTA4H* gene you have inherited, you could see an inflammatory response to TB that is 'too much,' 'too little,' or 'just right,'" according to Dr. Sarah Dunstan of Oxford Univer-sity's Vietnam unit, the coauthor of a study that appeared in February 2012.[40] Variations in the gene alter biological pathways to produce different levels of inflammation after infection, which could be crucial in determining who will contract the disease and who would benefit from steroids.

Even in psychiatry, pharmacogenetics is coming into early use to lessen adverse reactions to powerful antipsychotic drugs used to treat schizophrenia.

However, a ten-year review performed in 2007 is cautious: "These developments can be considered as successes, but the objectives of bringing pharmacogenetic and pharmacogenomic research into psychiatric clinical practice are far from being realized. Further development of genetic tests is required before the concept of tailored treatment can be applied to psychopharmatherapy."[41]

ONLY THE STUPID CANCERS?

These are some of the most positive findings in pharmacogenetics at the time of writing. Surprisingly, however, for a field prone to hype and overreporting of positive results,[42] there's also a lot more open controversy and criticism from fellow scientists and medics than you might expect to find. The former head of the American Society of Clinical Oncology, George Sledge, has gone so far as to declare that the only cancers that we've outwitted so far are the "stupid" ones.

What Sledge means is that up until now, oncology pharmacogenetics has mainly succeeded in developing drug treatments for cancers that have a single mutation in their genome. Herceptin is one example, Gleevec another—a highly effective drug that's turned the formerly fatal cancer myelogenous leukemia into a chronic disease. As Sledge says of Gleevec, "When it does fail, other drugs that work in the same way can put the cancer back into check. The reason is that the cancer is caused by a single rearrangement of DNA in blood cells. It turned out to be an easy target."[43] Myelogenous leukemia is a "stupid" cancer: it's only mastered one trick.

Outside oncology, one-trick (or sometimes two-trick) ponies also account for most of the successes so far in pharmacogenetics. If your response to tuberculosis infection depends on which variant of the *LPT4H* gene you've inherited, that's one example in which a single gene is crucial. The use of genetic testing to identify patients at heightened risk of bleeding from the anticoagulant drug warfarin likewise depends on the fact that only two genes account for most of the variation.[44]

But most diseases—unfortunately including some of the commonest and most lethal illnesses, such as lung and colon cancers—are cleverer than that. Some—like the fifty separate breast cancers that shared no more than three mutations[45]—are very Machiavellian indeed.

That's why the unfortunate bottom line, as the oncologists Lee and McLeod put it, is still that "most [cancer] drugs are not active in the majority of patients." The advances listed in the previous section—genuine though they are for those patients who do benefit from them—haven't yet changed that position radically. "Despite increasing experimental and clinical data on pharmacogenomics in oncology, an embarrassing gap between pharmacogenomic knowledge and clinical application still exists. . . . The efficacy rates are still inadequate in most cancers, in spite of the introduction of newer therapeutic options."[46]

In the words of the last person from whom you might expect to hear this admission—Allen Roses, the vice president of genetics at the pharmaceutical company GlaxoSmithKline—"Our drugs do not work on most patients."[47] Pharmacogenetics in oncology hasn't substantially altered that position as of yet. Cancer patients have every reason to want to hope, but it's still often hoping against hope. Comparing the efficacy of pharmacogenetic tests in eight different areas of cancer medicine, Lee and McLeod note that the predictive value of some tests, such as Innotecan efficacy for colorectal cancer, is no better than half—the same as tossing a coin.[48]

As we saw in chapter 2, the American Society of Clinical Oncology[49] accepts that genetic testing for personal cancer susceptibility is now a routine part of clinical care but warns that high-penetrance mutations (of the BRCA1 and BRCA2 variety) only cause a small minority of cancers, although women who inherit the unfavorable alleles have a very high probability of developing breast or ovarian cancer. One real advantage of personalized medicine is that these patients are benefitting from early screening to detect cancer at a less advanced stage. But even in cases linked to only one or two genes, genetic testing itself may be necessary but not sufficient for pharmacogenetic therapy—or indeed any therapy. Huntington's disease has been known since the early 1990s to be directly linked with the number of repetitions of one particular genetic marker, but it remains incurable, with recent clinical trials having proved unsuccessful.[50] Similarly, the genetic basis of cystic fibrosis has been known since the 1980s, but although treatment has improved, there's still no cure in sight.[51]

Trials in cancer pharmacogenetics additionally have to contend with an inherent paradox of personalization: the more unique or specific the proposed drug is to particular genetic subgroups of patients, the harder it becomes to find enough patients for statistically significant results. This profound problem makes some commentators skeptical that individualized drug therapy will be possible for most conditions any time in the foreseeable future.

The continuous discoveries, even today, of new surprises about our genome cause us to question reviews declaring that "personalized medicine is almost here" or that "individualized drug therapy will soon be a reality." . . . Numerous reasons exist to show that an "unequivocal genotype" or even an "unequivocal phenotype" is virtually impossible to achieve in current limited-size studies of human populations. This problem (of insufficiently stringent criteria) leads to a decrease in statistical power and, consequently, equivocal interpretation of most genotype-phenotype association studies. It remains unclear whether personalized medicine or individualized drug therapy will ever be achievable by means of DNA testing alone.[52]

Before we achieve anything like a paradigm shift in personalized medicine, we'll have to get a lot cleverer in treating those subtler but far more common cancers, the ones that aren't caused by one or two genes alone. "One danger of stupid cancer is that it makes us feel smarter than we are," Sledge concedes ruefully.[53] That overconfidence is obvious in many of the more exaggerated paeans to personalized medicine. Ironically, we can best advance the cause of personalized medicine by being more skeptical about what it can and can't do, avoiding the tragic fault of hubris.

THE WRONG TREATMENT FOR THE WRONG PATIENT AT THE WRONG TIME

The ethos of pharmacogenetics is "the right treatment for the right patient at the right time."[54] But the inevitable corollary is "the wrong treatment for the wrong patient at the wrong time," conceivably meaning "no treatment."

With kidney transplants, for example, it now seems feasible to use genetic markers to predict immune rejection versus survival.[55] The level of expression in the hematopoietic (blood-forming) cells can be used to forecast the incidence of graft versus host syndrome, which is crucial in rejection. As altruistic kidney donation symbolizes We Medicine at its noblest, would this use of personalized prediction in deciding who gets the donated kidney represent an unwelcome intrusion by Me Medicine? Or would it actually make better use of the donor's generous gesture? Supporters of Me Medicine might argue that it's both more efficient and more equitable to use advances in genetic sequencing to order to inject an element of rationality and justice.

How should we decide what probability of response to a particular pharmacogenetic treatment to use as a cutoff point? Should a patient get a drug even if there is only a 20 percent chance that she will respond? 10 percent? 1 percent? The standard response in decision-making theory or cost-benefit analysis[56] would be to rank patients according to the expected value of a treatment for them, obtained by multiplying the probability times the utility of the beneficial outcome. But we would also need to subtract the expected value of any *harmful* outcome, the iatrogenic harm inflicted by serious side effects.

However, the use of cost-benefit analysis is controversial because expected value is only one of several possible criteria for allocating scarce medical resources fairly.[57] Resource-allocation decisions are among the most serious issues in biomedical ethics—what Guido Calabresi and Philip Bobbitt call "tragic choices." There can be no universally happy outcome, however the protagonists act—even though some methods of allocation camouflage the element of tragedy better than others.[58]

Means of resource allocation are often subdivided into social and medical criteria. Social criteria in rationing treatment have rightly had a bad press since the 1960s, when the body that came to be called the Seattle "God Committee" allocated kidney dialysis on the basis of Boy Scout leadership, church membership, and income level.[59] Other social criteria, particularly age, seem to evoke much less hostility or even to seem commonsensical, on the argument that older people have had a "fair innings."[60]

Medical criteria can be subdivided into *prognosis* (the disputed resource should go to those with the best chance of recovery, using it with maximum productive efficiency) and *diagnosis* (no, it should go to those with the most pressing medical need, even if their chances of recovery aren't great).[61] Genetic matching falls into the first category, using prognosis as the guide—predicting immune rejection and allocating the kidney in such a way as to minimize the risk. But we generally favor the second category, gravity of diagnosis, for terminally ill patients. We think that the dying should receive more time and care than they "deserve" in terms of their poor prognosis. Some medical ethicists[62] argue that we should give greater weight to the desire for life in the patient with a shorter expectation of it, so that a terminally patient might actually be given preference in allocating a scarce resource, even though fewer total life-years are gained overall by such an allocation policy.

But what if we're pitting one terminally ill patient against another? Is it right to allocate the kidney by genetic matching in that case? Kidney transplantation isn't always the treatment of last resort—if there's still the option of dialysis, burdensome though that is—but "every four hours a patient awaiting a kidney will die."[63] Waiting lists for heart, liver, and lung transplants, which frequently *are* the treatment of last resort, likewise register high death rates. So there's an important question to be settled about justice, even if medical criteria prevail over social ones.

For terminally ill patients whose only hope is a transplant, rigorously applied genetic matching may deprive them of their only chance, although it improves the overall efficiency of organ allocation. Besides, you can't be absolutely certain that genetic markers would always predict the right recipient. Some kidney surgeons are skeptical that the science has really advanced that far.[64] They fear that their clinical judgment will be shunted aside in favor of supposedly objective but often inaccurate tests. In that case, you could argue, personalized medicine will have resulted in *less* control for clinicians and patients alike—the opposite of the usual promises.

Similarly, testing for a gene (*KRAS*) associated with colorectal cancer can partially predict that the patient is unlikely to respond to a particular therapy, especially if the tumor has already metastasized. It's easy to foresee a situation where a patient who wants this treatment, all the same, is denied it by his insurers or clinicians on medical grounds. Yet the sensitivity of *KRAS* testing is low: there are many people in the group with the "better" allele who don't respond to the treatment either.[65] So is it fair to favor them if treatment is expensive or scarce? You won't know until after the treatment decision is made whether they've responded or not.

In 2011, results were reported from a ten-year study of women undergoing chemotherapy for the type of breast tumor that *doesn't* respond to Herceptin—*HER-2*-negative cancer.[66] While *HER-2*-negative cancer has a better prognosis than its *HER-2*-positive counterpart, the tradeoff is the lack of other treatments than standard chemotherapy for the 90 percent of women in the negative camp.[67]

In the trial, 310 patients with *HER-2*-negative tumors were genetically tested for their chemotherapy resistance, which was traced to particular genetic markers. This genetic marker screen was then used on 198 new patients, to suggest whether they would be better off without chemotherapy—whether instead they should be offered an alternative, hormone-based therapy. Of the women who were predicted to respond well to chemotherapy, 92 percent survived three years without a relapse—an 18 percent lower rate of death than that for the group identified as genetically resistant.

It's this second group that concerns me—the genetically resistant. As a medical ethicist, I have aching doubts about dividing people into treatment/nontreatment camps based on a genetic test. If this ever became a routine basis for denying treatment to women who were predicted to be genetically unresponsive, I'd start to worry. Apart from the very real possibility of false positives and false negatives, there's the crucial question of whether the women classified as genetically resistant to chemotherapy might still want to take their chances on it but have their request denied solely on the basis of the adverse prediction. The project leader, Fraser Symmans, was cautious about taking any such radical step: "This result doesn't necessarily mean [that] women should abandon chemotherapy completely, but that they should consider an additional or alternative treatment."[68] But what if there is no alternative treatment?

Patients' enthusiasm for pharmacogenetics would take quite a hit if they saw it as a rationale for denying them therapy, but in an era of cost cutting, that's exactly what could happen. For example, a recent systematic review argued that warfarin management should depend almost entirely on cost effectiveness.[69] Even in a society such as Sweden, with its universal medical coverage and fabled sense of solidarity, commentators predict that pharmacogenetics could be used not to enhance social justice but to rein in patient demand.[70]

If the concern is using public resources fairly, justice seems to demand that the diseases that affect the greatest number of people should receive the highest priority. But what happens to the others? Pharmacogenetic analysis reveals not only the old problem of "orphan diseases"—those that only affect small minorities of patients—but also "orphan patients," those unfortunates who are genetically "programmed" to be unresponsive to particular treatments.

It's conceivable that drugs might be developed for small populations of patients, particularly if industry is offered incentives under the Orphan Drug Act of 1983, but only if drug firms can charge high prices to make up for the lack of numbers. Just how high those prices can go was exemplified recently in the case of a pharmacogenetic treatment for some variants of cystic fibrosis, a combination regime of Kalydeco (ivacaftor) and another drug called VX-809. A year's Kalydeco treatment is currently priced at $294,000.[71] This price seems particularly exorbitant given that cystic fibrosis, the most common genetically linked disease in whites, isn't even an "orphan" condition. What accounts for the high price tag?

Kalydeco, produced by Vertex Pharmaceuticals, was licensed by the FDA in January 2012 on the basis that it had produced significant and sustained improvement in lung function in two trials. However, it was only effective in the 4 percent of cystic fibrosis patients who have the *G551D* mutation of the gene coding for the disease—about 1,200 of the thirty thousand Americans diagnosed with cystic fibrosis. Vertex is hoping to reach a wider range of cystic fibrosis patients with the combination of Kalydeco and VX-809, which can have positive effects on lung function in patients with the more common genetic variant. Yet it looks as if the company is still basing its pricing policy on the assumption that even the combination will only be marketable to a limited number of patients.

With patents expiring on many traditional "blockbuster" drugs of the one-size-fits-all variety, the pharmaceutical industry is hoping that tailored therapy will allow them to claim "a relatively large share of a smaller pie," as Lilly's CEO Sidney Taurel put it.[72]

> The traditional objective of pharmaceutical companies has been to achieve "blockbusters," new chemical entities satisfying (or creating) unmet medical need in a substantial (and preferably chronic) patient population. Pharmacogenetics is more likely to produce a number of "mini-busters" (products with a relatively high cost for low volumes that are tailored to the needs of well defined, relatively small groups).[73]

The isolation of being the wrong patient with the wrong disease at the wrong time represents the ultimate extreme of the turn from social determinants of disease to individual ones. In the nineteenth century, rich and poor alike were vulnerable to epidemics such as cholera, smallpox, and typhus. The health measures that improved life expectancy for everyone, those familiar We Medicine heroes like sanitation and vaccination, served both social justice and individual well-being. But when infectious disease was replaced by cancer and cardiovascular disease as the main causes of mortality, illness was individualized—as was medicine's response. Pharmacogenetics takes that atomistic logic one crucial step further. The question is whether it's a step too far.

There's no incentive for the medically better off—those who *will* receive the treatments allocated by pharmacogenetic standards—to care about the "orphan patients." They'd have to be persuaded by what you might call a "genetically enhanced" version of the argument that Tom Wilkinson and Kate Pickett champion in their influential 2010 book *The Spirit Level*: equality really does produce better health outcomes for everyone, not just the poor. But what is equality in pharmacogenetic terms?

Aristotle famously defined justice as "treating equals equally" and unequals unequally.[74] Unlike the contentious social categories that Aristotle himself used to define who counts as equal—women don't, for example—pharmacogenetics appears to offer a scientific and objective standard of who would most benefit from limited medical resources. Treating those who are genetically less likely to benefit just wastes the resource, whether it's a kidney or a drug treatment. Those, at least, are the arguments that the genetic "elite" could use against a leveling argument of Wilkinson and Pickett's sort.

Balancing individuals' and patient groups' need for medical resources has to be done in any healthcare system, whether socialized or market driven. The unanswered question is whether pharmacogenetics will produce a more equitable model. There are uncertainties about both the state of the science and whose interests are served.[75] That's not just a technical quibble: it goes to the heart of society's deepest divisions, as the story of BiDil shows—a drug that initially appealed because it seemed to rectify past injustices to African Americans.

BIDIL: BOTTLED "PHARMACOETHNICITY"

BiDil (pronounced "bye-dill") was the first race-specific drug ever approved by the FDA, intended for a single ethnic group: African Americans suffering from congestive heart failure. Strictly speaking, it's not a pharmacogenetic drug—because the target is self-identified African Americans, not those singled out as such by genetic sequencing, and because in any case race is not a scientifically valid genetic category. But the rise and subsequent fall of BiDil tell us quite a lot about the commercial and medical intricacies of personalized pharmaceuticals. That's certainly how the drug was heralded: as "a step towards the promise of personalized medicine," in the words of the Food and Drug Agency when it announced its approval.[76]

Some commentators think that drug companies may be tempted to find shortcuts to personalization, and race is no doubt the most obvious. "Scientists and entrepreneurs see race as an avenue for quickly translating the embryonic science of personalized medicine into marketable products." BiDil represented "pharmacoethnicity" in a bottle—"color-coded pills."[77] The drug was promoted as *both* Me Medicine (deliberately distinguished from the one-size-fits-all model) *and* We Medicine (equally consciously appealing to black identity politics).

Because ethnic minorities[78] are growing in relative population size, and because they suffer disproportionately higher rates of heart disease, diabetes, and other chronic illness, the pharmaceutical industry has long been "making it a high priority to cultivate relationships with ethnic consumers," as the fifth annual Multicultural Pharmaceutical Marketing conference announcement put it in 2004.[79] While some might argue that this is a form of overdue restorative justice, little about the marketing of BiDil deserves that name. The brand-name drug cost between four to seven times as much as the two generics that composed it, but it was marketed to one of the poorest patient groups in the United States.[80] Its high price, at $1.80 per pill, or nearly $4,000 a year for the recommended six pills daily,[81] placed it beyond the reach of many patients—disproportionately people of color—who would have benefited from treatment with the two generics separately.

Nor was BiDil actually developed with African Americans in mind, although the biotechnology industry spokesman Michael Warner encouraged his audience to think so when he claimed, "BiDil is the first time, the highest profile time, the model of 'let's identify a target population and let's develop a drug for that population' has been pursued."[82] What really happened is this: after trials on a general multiethnic patient group had failed, "data dredging" was used to construct a scientifically unsubstantiated case for restricting the drug to blacks. If that sounds implausibly cynical, let's look at the backstory.[83]

In the 1970s, a cardiologist called Jay Cohn developed a treatment using chemical vasodilators for congestive heart failure—half of whose victims die within five years of diagnosis if their condition is left untreated. At the time, there was in fact no other effective treatment, so Cohn's beneficent motives are unquestionable. Cohn thought that relaxing the blood vessels might help hearts weakened by high blood pressure, infection, or heart attack to circulate sufficient volumes of blood. The two vasodilators with which he experimented were the generic drugs hydralazine hydrochloride and isosorbide dinitrate, which could both be taken orally. These are the two ingredients in BiDil: the only difference is that they're combined in a single pill, whereas originally they had to be taken separately.

With funding from the Veterans Administration, Cohn and his colleagues conducted two five-year clinical trials using the hydralazine hydrochloride and isosorbide dinitrate (H-I) combination, taken separately, on a total of 572 men—both white and black. In the first trial, compared against a placebo and a drug for high blood pressure, the H-I combination fared well. But in the second, it was found less effective than an angiotensin-converting enzyme

(ACE) inhibitor called enlapril, which became the treatment of choice for most heart failure patients.

So far, so scientific, but at this stage commercial considerations began to outweigh medical ones—or at least to conflict with them. Cohn wanted to test what would happen if he treated patients with both the H-I combination and enlapril, but his VA funding had run out. Because the two ingredients were cheaply available generics, no commercial company was interested in sponsoring the trial, since no patentable drug would result. At this point, in 1989, Cohn opted to combine the two drugs into a new fixed-dose pill called BiDil and to file for a patent. The patent application said nothing about the race of the patients for whom it was intended.

Although Cohn got his patent, which allowed him to interest a small North Carolina drug company called Medco in taking up a license on the rights, he failed in his quest for FDA approval of BiDil. Rather than running a new trial, he'd simply reinterpreted the old data, which the FDA said didn't meet the tests of statistical significance. In any case, the FDA reported, the results of the second trial had already shown that the H-I combination was less effective than the ACE inhibitor. When the FDA turned Cohn down in 1997, Medco dropped the drug.

At this point, Cohn had no commercial backer, no FDA approval, and a drug whose patent was due to expire in under ten years. That's when he decided to turn BiDil—which hadn't been presented to either the FDA or the patent commission as a racially targeted drug—into a therapy for African Americans. Together with another lead researcher on his clinical trials, Peter Carson, Cohn returned again to the old data and divided it into racial groups.

Cohn and Carson published a paper in 1999 reporting that "the H-I combination appears to be particularly effective in prolonging survival in black patients."[84] This conclusion was based on only forty-nine black and 132 white patients who had received the H-I combination in the first trial and 109 blacks versus 282 whites in the second. A properly conducted trial to determine racial difference in response would have needed to start from scratch, to enroll larger numbers of both races and to control for other variables that might affect differential response.

For example, blacks might possibly have higher background rates of hypertension or obesity from poorer living circumstances.[85] One in three African Americans has hypertension, compared to one in five whites.[86] That's probably attributable to more stressful environmental factors (although the Oprah show reinforced the genetic explanations when the star suggested that slaves with a genetically based tolerance for retaining salt survived the Middle Passage more readily, because they were less susceptible to dying of thirst, and that their descendants were genetically prone to hypertension as a result.)

In the same year, 1999, Cohn negotiated with a new commercial backer, the Boston-area firm Nitromed. Together with Carson, he filed for a new patent on BiDil to replace the one that was set to expire in 2007. Although the drug hadn't changed, the patent application's wording had. Cohn and Carson now claimed that "the present invention provides methods for treating and preventing mortality associated with heart failure in African American patients."[87] With patent in hand and commercial backing behind him, Cohn was in a stronger position to approach the FDA again—this time with a drug purportedly perfect for African Americans. And this time the FDA said yes, it might give its approval—provided that Nitromed could raise venture capital for a clinical trial specifically on blacks.

Nitromed raised that money—a testament to the strategic appeal of "niche marketing," with race as a niche like any other—in a market looking for new investments after the collapse of the dotcom boom.[88] Backed up by $31.4 million in private finance and the approval of the Association of Black Cardiologists, the African-American Heart Failure Trial (A-HeFT) duly opened for business in 2001. It enrolled 1,050 self-identified African American men suffering from advanced heart failure —this time along with female patients, the largest number of black women ever to participate in a clinical trial. Half of all the research subjects received BiDil as well as standard treatment, half the standard treatment alone. That sounds like correct practice for running clinical trials, but note one thing: there was no other ethnic group for comparison.

In other words, if the treatment worked well for blacks—and it turned out that it did, so well, in fact, that the trial was called off halfway through—it might have worked equally well for other people of color and whites too. As Jonathan Kahn testified at the FDA hearing, "Most drugs on the market today were approved by the FDA based on trials conducted almost exclusively in white patients, but those drugs are not designated as white drugs, and rightly so."[89] If every drug tested on whites alone were only prescribed for whites, people of color would have a greatly limited choice of treatments. Cohn himself has frequently repeated that he prescribes the combined ingredients in BiDil in color-blind fashion, to white and black patients alike. But "Nitromed had a financial disincentive for finding that BiDil worked regardless of race— its patent (and market monopoly) applied only to its use by African American patients."[90]

What had persuaded the Association of Black Cardiologists to lend their name to this trial? They weren't alone: the Black Congressional Caucus, the National Medical Association, the National Association for the Advancement of Colored People (NAACP), and the National Minority Health Month Federation all joined them in urging the FDA to give BiDil the green light for African American patients. Juan Cofield of the New England NAACP branch told his members in 2006: "I would like to see the name BiDil [become] as common in our community as Viagra is to the general public."[91]

These organizations were probably won over by the much-bruited statistical claim that blacks were twice as likely to die from congestive heart failure as whites because they had a shortage of nitric oxide. BiDil's second ingredient, isosorbide dinitrate, increases the level of nitric oxide in the blood. But the statistical claim was wrong: it came from a 1981 study that had subsequently been discredited. Later studies showed that there was a racial difference in congestive heart disease deaths, but only at a ratio of about 1.1 to 1, blacks to whites—not double the numbers.[92] That difference was small enough to be caused by other, environmental factors disadvantaging blacks.[93] Those factors could have been tackled by public health programs—We Medicine— but instead individual African American patients were being targeted as purchasers of a drug personalized by race—Me Medicine.

Yet "health inequities—which are caused by social injustice, not genetic dissimilarity—cannot be fixed by color-coded pills."[94] Of twenty-nine medicines claimed to differ in power across racial or ethnic groups, a study by molecular biologists found that only a single drug (beta blockers for hypertension, which are more effective in Europeans than in African Americans) demonstrated any genuine statistical difference.[95] An influential report from the Center for American Progress accuses racially profiled drugs of diverting attention from the greater healthcare disadvantages suffered overall by blacks. Promoting racially profiled drugs is also insensitive, the authors assert, to the long and troubled history of race in medical research— the sorry tale that includes Henrietta Lacks and the Tuskegee syphilis "studies."[96]

Some genetic conditions, such as Tay-Sachs and Canavan's disease, are genuinely linked to particular ethnic groups, such as Ashkenazi Jews. Yet typically they aren't exclusive to only one ethnicity: Tay-Sachs disease also occurs among French Canadians, Louisiana Cajuns, and Pennsylvania Dutch. Sickle cell disease, generally associated with African Americans or Afro-Caribbeans, is found in many populations exposed to malaria, against which a single allele of the gene confers protection (although two alleles produce the disease). The highest rate of the disease is actually in a Greek population, whereas Kenyan ethnic groups living at mosquito-free altitudes have low rates.[97]

Back in the bad old days before modern science, a particularly risible theory invoked a negative-weight substance, phlogiston, to explain why matter loses weight during the process of combustion. There are no negative-weight elements, of course, but at least the process of combustion really exists. That's more than can be said for the "fact" that blacks were twice as likely as whites to die of congestive heart failure. So you could say that phlogiston theory is only half as unscientific as the reasoning behind BiDil. The combustibly unscientific category of race was being used to account for a nonexistent difference. "Pooling people in race silos is akin to zoologists grouping raccoons, tigers and okapis on the grounds that they are all stripey."[98] But at least raccoons, tigers, and okapis have real stripes.

As Kahn puts it (a secondary credit goes to Malvolio in *Twelfth Night*): "Some drugs may be born ethnic, others may achieve ethnicity, but BiDil had ethnicity thrust upon it."[99] In the end, however, personalization by ethnicity wasn't enough to sell the drug. With sales languishing, Nitromed's share price had bellyflopped from $23 to $3 by 2007. Even the company's offer to establish a financial assistance program, NitroMed Cares, couldn't turn BiDil around.

Insurers refused to pay the extra $3,000 a year that combining two generics into one branded drug would cost. Likewise, Part D Medicare and Medicaid recipients found that the drug wasn't covered. While pharmaceutical industry bodies tried to brand that decision as racist—since blacks were heavily represented among patients who depended on Medicare and Medicaid—focus-group studies found that African Americans themselves were suspicious of drugs marketed as specially for them.[100] If the drugs worked, some of them reasoned, whites would have grabbed them too.

By 2007, the Association of Black Cardiologists had retracted its initial support for the drug. But the reason why BiDil was withdrawn from the market in 2010 was not primarily because it was a medical failure (it was medically neutral, no more or less effective than the two generic drugs in combination) but because it was a commercial flop. There's a certain justice in that: it wasn't launched because it was a medical breakthrough but because the business exigencies of patenting and niche marketing initially made it look good. Yet there's also a nagging injustice: the clinical trials did show that BiDil was highly effective—although expensive—in combination with other standard therapies. But by stratifying the drug's audience by race, Nitromed had effectively deprived both blacks and whites of its potential.

It would be naive to think that pharmacogenetics could be any more exempt from commercialization than other aspects of modern biotechnology. BiDil may be a particularly egregious example, but the history of personalized drug regimes also evokes some disconcerting echoes of the way in which the tobacco companies apparently hoped that genetic screening would identify susceptible individuals, so that "the rest could be allowed to puff away contentedly." (While smoking is more dangerous for some individuals than others, for example, people with a deficiency of alpha-1 anti-trypsin, of course there is no one for whom it is safe.) Likewise, Helen Wallace alleges, drug companies aimed to weed out patients who were genetically susceptible to adverse drug reactions, so that the profitable pattern of megadose prescribing of blockbuster drugs could continue unchecked.[101]

Before we leave the BiDil case, there's one final important lesson to draw from it. Patentability drove BiDil's history: both the pressure to extend the old patent's life by taking out a new one and Cohn's solution of resurrecting the drug as stratified by race. Without a good intellectual-property prognosis—a long and happy life for the patent—there was no chance of interesting a commercial backer, no way of running further clinical trials, and no hope of FDA approval. In the BiDil story, patenting plays the roles of both the good fairy and the bad fairy: it gives the drug life, but with what, if you don't mind a bit of hyperbole, might be called a curse.

In the case of Herceptin, patenting acted much more like the bad fairy. Although Genentech did little to encourage the development of the drug,[102] its patent on *both* the drug and the *HER-2* gene *itself* drove the price of Herceptin beyond what many insurers and patients could afford. A monopoly on the gene meant that no competitor could invent a cheaper drug, discouraging valid medical research that could benefit patients.[103] The UK National Health Service had to restrict Herceptin's use for cost reasons, until it was forced to retreat from that policy by public pressure.

Given that one in five human genes is now the subject of a patent[104] and that patents are frequently the most valuable item in a biotechnology company's portfolio,[105] patents may well restrict the affordability of other pharmacogenetic drugs. Herceptin is the poster child in more ways than one. If other patents on genes themselves are enforced as rigorously as that one was, pharmacogenetics might actually *increase costs* and *cut choice for the neediest patients*—the exact opposite of the hopes held out for it. In increasingly hard times, payers may simply not be able to afford the elevated prices, made possible by a monopoly patent on the genetic marker involved, that helped push global sales of Herceptin to the level of $5.4 billion in 2010.[106]

THE FOUR HYPOTHESES REVISITED

In chapter 1, I offered four possible explanations for the rise of personalized medicine: threat, narcissism, corporate interests, and choice. Let's sum up how these four factors might apply to pharmacogenetics.

First, *threat* is certainly relevant (contamination less so). Pharmacogenetics is disproportionately (although certainly not exclusively) about cancer treatment. Susan Sontag's case, one among hundreds of thousands, illustrates the dual threats of the increased incidence of cancer and of harm from the very treatments used to combat it. Both combine to make cancer the most feared disease.[107] The sense that you're damned if you do, damned if you don't treat cancer, and the disease's portrayal as almost inevitable,[108] merge to create something very like desperation in the popular view of cancer. (However, the emphasis on cancer in pharmacogenetics isn't driven by the popular sense that cancer and its treatment are both appalling threats, rather by the fact that more is understood about the molecular pathology of cancer than about other common diseases.)

Pharmacogenetics appeals strongly to this sense that nothing can be done against the threat of cancer, not least when what could be done—old-style scorched-earth treatment—is itself an iatrogenic threat. The St. Louis leukemia patient whose genome was sequenced was able to choose targeted chemotherapy rather than a more invasive treatment, bone-marrow transplant. Within chemotherapy, those patients who are genetically programmed to respond well to a particular drug can potentially be offered a dosage that won't entail such poisonous side effects. Susan Sontag might have been one of them, although we'll never know.

Second, precisely because the threat of iatrogenic harm is so real in cancer care and because much pharmacogenetic research in that area concerns lessening that threat, locating the drive toward personalized medicine in national *narcissism* seems inappropriate for pharmacogenetics. True, BiDil was marketed on the back of the Black Pride movement, but that's not narcissism as

Twenge and Campbell use the term. In any case, BiDil failed as a commercial venture.

However, there is one sense—albeit far from Putnam's own—in which pharmacogenetics is more about "bowling alone" than is generally recognized: the situation in which seriously ill patients can't get the pharmacogenetic drugs they need, for economic reasons. That leads into the third hypothesis, *corporate interests and neoliberalism*.

In cancers driven by a number of different genetic pathways, different patients may need different regimens of combinations of drugs. With drugs required by smaller-size patient groups, it may not be economic for drug companies to produce every drug required for the regimen of any particular patient. For several years, insurers and health maintenance organizations have been cutting down on the number of drugs they cover, restricting formularies and encouraging therapeutic substitution.[109]

We're also witnessing the stratification of patients into genomic niches, as genetically predisposed or genetically resistant. Individualization through genetic stratification may well have the consequence that patients who fall into the "wrong" genetic category—because they're less genetically likely to respond to a particular therapy—will find themselves denied treatment. This seems all too plausible given the economic and political climate of austerity in which pharmacogenetics is being developed. In that climate, patients may also be denied pharmacogenetic treatments known to be effective if monopoly gene patenting pushes the cost too high for healthcare payers to afford, as occurred with the initial NHS response in the Herceptin case.

From the drug companies' point of view, it's been big blockbuster drugs that have traditionally been the ones printing money. Unless a stratified patient group is large (or wealthy) enough to constitute a niche market, it won't necessarily be in the drug companies' interests to tailor medicines too narrowly. Alternatively, they may pursue a strategy of high price increases for personalized drugs: the pricing of a group of oral oncolytic (antitumor) drugs, including Gleevec, has gone up by over 76 percent since 2006.[110] We saw the same strategy at work in the example of Kalydeco for a type of cystic fibrosis, where the price of the drug was set extremely high, presumably to offset the small genetically tailored market.

On the other hand, some patent experts predict that sequencing the genome will provide an enormous stimulus to the pharmaceutical industry, logarithmically expanding targets for drugs from about five hundred pre-HGP to nearly ten thousand now.[111] Next-generation personalized drugs are already being pursued, including Xalkori, which was developed with a small group of patients whose lung cancers had a particular mutation, making them possibly susceptible to treatment with an inhibitor for a protein called ALK. The trials were rapidly successful, more quickly than trials with a randomly selected group of patients would have been, with the drug derived from them (Xalkori) being made available at a price of $9,600 per month.[112] Although that's less than the cost of Kalydeco, the comparatively high price is again driven by the small size of the potential market: the total target population for the drug is expected to be fewer than ten thousand patients.[113]

In fact, Xalkori is cited in an article in *Nature Reviews: Drug Discovery* as a prime example of the way in which "biomarker-driven agents [drugs] necessarily lower the patient population."[114] Drug companies accustomed to relying on mass-market blockbusters will have to develop a different business plan if pharmacogenetics is to become commercially viable. So far, that strategy seems to ignore the risk that patients or insurers may simply refuse to pay high prices for genetic niche drugs: instead, "premium pricing" is seen in the industry as a risk-reducing strategy.

> Although these [pharmacogenetic] agents are important advances in treating patients, they introduce an element of unfamiliar commercial risk for drug developers. Firms have engaged in several strategies to combat this risk: foremost among them has been a steady increase in the prices of pharmaceutical agents. . . . Overall, premium pricing is seen as an essential risk-mitigation strategy in the development of agents targeting novel biomarkers.[115]

This bears out my contention about the problematic economic consequences of pharmacogenically stratified drug production and the consequent likelihood that some drugs may either not be produced or be priced at high levels that insurers or patients can't afford. "To make these price increases sustainable, manufacturers must be able to justify the increased cost needs to payers and health-care providers."[116] If the payers won't pay, then the drug may simply not be available, because it's uneconomic for the manufacturers to make it.

So corporate interests are highly relevant to the impact of pharmacogenetics: no matter how effective the science is, this form of personalized medicine won't be there for some patients if corporations can't afford to produce the drugs at a price patients and insurers can afford to buy. The alternative might be for particular patient populations to seek government subsidies, which the drug companies would probably welcome. Public funding for private companies at the risky stage of drug development might be acceptable to neoliberal governments, but especially in economic hard times, will they be able to issue a blank check of ongoing subsidies?

Finally, patients can't choose expensive pharmacogenetic drugs that their insurers or healthcare system won't pay for—not unless they have the means to foot the bill themselves. Yet implicitly, patients have been encouraged to think that pharmacogenetics will increase their *choices* and their *autonomy*: that they'll be the right patient, as in "the right drug for the right patient at the right time." But if there are right patients, there also have to be wrong patients, whose options will actually be narrowed: perhaps those in genetically identified "orphan" treatment groups, as well as those with the more traditional problem of having minority "orphan" diseases. Even for the fortunate, some clinicians fear that pharmacogenetics might mean that treatment algorithms—paint-by-numbers medicine—could come to predominate over patient-centered consultation.[117]

On the other hand, some treatment plans tailored to the patient's genomic sequencing do give genuinely meaningful choices. Women undergoing IVF would welcome an opportunity to take a lesser dosage of hormone stimulation if they knew that they were likely to respond readily to it. Pharmacogenetics could offer them that choice—although another low-dosage option is already available to them in the form of clinically effective "mild" stimulation regimes, or even no stimulation at all.[118] Cost considerations would actually favor pharmacogenetics in that case, since the lower-dosage or natural regimes would be cheaper.

In a very important sense, however, choice is irrelevant and even offensive in pharmacogenetics, particularly in oncology. Patients don't have cancer by choice, unless you take an unsympathetic view about people with cancers traceable to "lifestyle choices" such as smoking. That seems unduly harsh to me.

Pharmacogenetics also throws up some odd ironies about choice's partner, *individualism*. In biological terms, that uniqueness that we all treasure so much would actually lead to the unlikely and economically untenable extreme of one patient, one drug. That's a deliberate exaggeration, of course, intended as a thought experiment, but the logic of individualism is genuinely problematic when it comes to pharmacogenetics.

Just as we're all so unique—even identical twins are subject to epigenetic modifications as they go through life—so too are the mutations in our cancers (at least the so-called smart ones). Yet here's the tension: the more intensely personalized pharmacogenomic research gets, the more likely it is to fail, and not just for economic reasons. To repeat a clinical assessment that we encountered earlier:

> The continuous discoveries of new surprises about the genome call into question the claim that personalized medicine is almost here, or that individualized drug therapy will soon be a reality. In fact, it probably never will be, or at least not by DNA testing alone, because most genotype-phenotype associated studies are hampered by limited size and therefore decrease in statistical power.[119]

One last anecdote: some of the most promising research in cancer prevention actually comes not from the complexities and costs of individually tailored drugs but from simple, cheap, and comparatively safe one-size-fits-all drugs, even for genetically caused conditions. In October 2011, a UK team found that a daily 600 mg dose of aspirin resulted in a 63 percent reduction in the number of colorectal cancers in patients with a hereditary disease called Lynch syndrome. This genetic condition increases the risk of colorectal and uterine cancer in about 2 to 7 percent of the population, by affecting genes responsible for detecting and repairing DNA damage.[120] Every one of the 861 people with this syndrome in the trial got the same dosage of the same simple drug against the same threat. It worked.

4

"YOUR BIRTH DAY GIFT"

Banking Cord Blood

THE PROMISE OF "ONE PATIENT, ONE DRUG" TURNED OUT TO BE A logical and logistical impossibility in the case of pharmacogenetics, but there is another kind of personalized medicine that claims that such a treatment is perfectly feasible. What's more, the "drug" is a form of stem cells in the patient's own blood. Yet it wasn't manufactured by the patient's body but by another person altogether, who gave it to the patient as "your birth day gift." (Yes, "birth day," not "birthday.")

What's the answer to this riddle? For over twenty years, it's been known that umbilical cord blood, taken during the baby's birth, contains hematopoietic (blood-making) cells. These cells are similar to but fewer in number than those found in bone marrow, although they are more flexible in their potential. The first instance of their use in clinical medicine was to treat a baby's older sibling who had Fanconi's anemia[1]—that is, they were used for We Medicine purposes, on someone else's behalf.

This usage, along with similar uses of hematopoietic stem cells for transplantation to someone else outside the family, has become well established. More than ten thousand patients worldwide have received such transplants,[2] with roughly 360,000 cord blood units stored internationally in public banks.[3] These banks don't charge parents a fee for collection and storage, although in the United States they do charge a fee to the recipient's insurance company. Importantly, they will also return the cord blood unit to the parents in case of need, assuming that it's still in the inventory.

It's from one of these banks that the phrase "your birth day gift" comes: the Anthony Nolan Trust, a not-for-profit organization, uses it in a leaflet to encourage expectant mothers to donate cord blood.[4] In 2011, the United Kingdom announced an initiative to double the number of units stored in its public bank to 35,000,[5] while France—never to be outdone by its *ancien ennemi*—pledged fifty thousand units for its own bank. Early in the following year, the UK Department of Health announced that it was granting £4 million to the public bank (NHS Blood and Transplant), matching the French figure. Even this number would still only serve 85 percent of the UK patients who require a cord blood transplant.[6]

Perhaps surprisingly, the United States already beats both the United Kingdom and France in this particular form of *fraternité*. In its twenty years of operation, the not-for-profit New York Blood Center alone has nearly amassed the French target number, with about 49,000 units available in its National Cord Blood Program. In 2007, the U.S. Health Resources and Services Administration, part of the Department of Health and Human Services, allocated $32 million to widen the collection in the national program further, although the New York Center is already just one of many. Overall, the United States has by far the highest number of public banks in the world. In this respect, at least, Americans haven't actually fallen prey to "bowling alone": it's a powerful practical testimony to civic-minded community.

More recently, however, cord blood has also become a much-touted part of Me Medicine, with parents being urged by private firms to bank cord blood for the baby's exclusive private use. While cord blood isn't "manufactured" by the baby but rather by the mother, the hope is that these cells could become a "spare-parts kit" for the baby because of tissue compatibility. With about nine hundred thousand cord blood units banked with for-profit firms,[7] there are now nearly three units stored privately for every one in a public bank. Private cord blood banks are springing up all over the world, even in India, where seven private banks number some twenty thousand samples among them.[8]

In the United States, private firms typically charge between $1,500 and $2,000 up front, plus an annual fee of $90 to $200 for storing the blood, or between $3,600 to $4,100 in total.[9] Storage is supposedly a form of biological insurance—although the practice on a routine basis is discouraged by the American Academy of Pediatrics, the American Medical Association, the

American College of Obstetricians and Gynecologists, and the American Society for Blood and Marrow Transplantation. Equivalent British professional bodies, including the Royal College of Obstetricians and Gynaecologists, have also issued guidelines recommending against routine collection of cord blood. Meanwhile in France, Belgium, and Italy, purely private cord blood banks are actually illegal, although as we've seen in the French case, public banking is strongly encouraged.

What Is Umbilical Cord Blood, and How Is It Collected?

Many people think that childbirth ends when the baby is born, but there's still a final stage of delivery to go: the expulsion of the placenta, to which the baby is connected by the umbilical cord. In traditional "expectant management" birth, the baby remains attached to the umbilical cord, and blood continues to pulse between the bodies of mother and child. The placenta is usually delivered within thirty minutes to one hour and is then separated from the umbilical cord.

In modern "active management" of the final stage, however, oxytoxic drugs are administered to hasten the separation of the placenta from the uterus. The baby takes a few breaths, the cord is clamped and cut soon thereafter, and the placenta is delivered by gentle pulling on the cord.

Cord blood can either be collected while the placenta remains attached to the uterine wall (*in utero*) or after the placenta has been delivered (*ex utero*). Particularly with the *in utero* method, much of the blood that would naturally flow to the baby will be diverted for storage, and that's been criticized—as has the practice of early clamping, which many obstetricians regard as dangerous. So cord blood isn't just a waste product that would otherwise be discarded, as is often claimed by cord blood banks. It requires active and controversial obstetrical interventions.

Cord blood is equally or even more effective for transplantation than bone marrow, since the cells are at an earlier stage of development and more "plastic": "Because they proliferate rapidly, the stem cells in a single unit of cord blood can reconstitute the entire hematopoietic system."[10] Lymphocytes in cord blood are more immunologically "naive" than those from an older donor and thus less likely to react against the recipient's immune system.[11] Nor does taking cord blood require as invasive a procedure as bone-marrow collection (although it's certainly not without its risks, which will be detailed later in this chapter).

Yet cord blood transplantation works best if it's *not* done with the patient's own cord blood but with somebody else's. Surprisingly and counterintuitively, autologous (one's own) blood is not only less effective than allogeneic (another's) blood; it may actually be dangerous, if the illness requiring a blood transplant is something like a genetically based leukemia, where the harmful mutation is present in the blood from birth.

So here's a striking example where We beats Me hands down. Public banking makes the most *medical* sense—as well as the most *ethical* sense, in the opinion of all the major professional bodies and many experts.[12] When well-intentioned parents bank cord blood privately, it removes another unit from the public supply—potentially harming everyone in the end, including their own baby. In the extreme case, if the private bank goes out of business, the cord blood is lost to the parents too. Me Medicine, in this guise, isn't even good for the individual, let alone the collective.

As in "the tragedy of the commons,"[13] when too many people use a public resource or too few contribute to it, everyone suffers—although it's rational for each individual to be a free rider. Yet there's nothing free about private banking: parents pay substantial sums, even though if their local hospital offers the service, they could bank the blood for nothing—and still have access to their own baby's blood in an emergency if it remains in the repository. So unlike the situation in the tragedy of the commons, it's actually *not* rational to bank blood privately. But while there is substantial evidence that Me Medicine in the form of private cord blood banking is bad for the collective *and* risky for the baby (and possibly the mother as well), it's not altogether clear that We Medicine—public banking—avoids all risks, as we'll see later in this chapter.

Nor is the division between public and private entirely straightforward these days, with public banks funding themselves by selling cord blood units to other banks on a global scale.[14] Others are collaborating with for-profit banks: for example, the partnership between Viacord; the National Heart, Lung, and Blood Institute; and the sibling-donor registry at the Children's Hospital of Oakland. Even France, with its long-standing commitment to free and anonymous donation of all tissue, is resorting to a public partnership with a private hospital group in order to meet its target of fifty thousand units.[15] Meanwhile, one UK-based private cord blood bank offers to donate 80 percent of the blood to a public bank "as a way to break down the public-private divide which has characterized the cord blood banking sector to date."[16] However, although parents pay less to store the 20 percent than they would to retain 100 percent of the blood, they're still paying hard cash to be altruistic.

The personalized technologies dealt with in the previous two chapters are genuinely beneficial at best—many developments in pharmacogenetics, for example—and ineffective at worst—if critics of retail genetic tests are correct. But here we have an example of Me Medicine which has been condemned by many medical professionals, on the one hand, as *actively bad* for society, the baby, and the mother, while promoted by its backers as life saving. What's the truth behind the riddle?

THE EVIDENCE BASE: BENEFITS

Banking your newborn's umbilical cord blood stem cells can be one of the smartest, farthest-reaching decisions you can make when it comes to the health and wellness of your newborn, and to your own peace of mind. Knowing you've taken steps to safely store these incredible, life-affirming *cells*—and that they're just a phone call away—can put any parent's mind at ease.[17]

This declaration about the benefits of banking your baby's cord blood is typical of private banks' publicity. I could have chosen any number of other advertisements, some far more assertive, but that would be taking aim at a straw man. What they have in common is the claim that cord blood can serve as a personal spare-parts kit for the baby. Although cord blood only contains hematopoietic cells in very low frequencies, expansion techniques being developed through stem cell regenerative medicine may be able to increase their number and to make use of their transformative potential. And since the cells are tissue matched to the child, this reasoning runs, there would be no immune rejection problem.

The Pacificord site goes on to declare, "Stem cells represent a new era in medicine, thanks [to] their uncanny ability to regenerate themselves into cells that form all other tissues and organs found in our bodies." Over seventy diseases have been cured or treated by cord blood stem cells, it's claimed, including heart disease, muscular dystrophy, Parkinson's disease, Alzheimer's disease, leukemia, multiple myeloma, osteoporosis, sickle cell disease, and systemic lupus. Many other private cord blood banks make similar assertions in their promotional literature: "like freezing a spare immune system," "unimaginable possibilities," "a miracle of nature . . . only available once in a lifetime."[18] Nor have these claims fallen on deaf ears: the great majority of pregnant women surveyed in one New York clinic believed that cord blood could already be used to treat Alzheimer's disease, Parkinson's disease, and spinal injury, which is not the case.[19]

There's a common thread here with the inflated medical claims made in other areas of Me Medicine—and with people's propensity to believe them: the conviction that the Human Genome Project would quickly lead to cures for everyday diseases, for example. The lure of imminent cures is all the more compelling when it's our children's health at stake. Tinged with the glamour of the stem cell technologies, cord blood banking apparently offers our children the proverbial elixir of youth. As Catherine Waldby and Robert Mitchell put it: "[It] allows them to live in a double biological time. The body will age and change, lose its self-renewing power and succumb to illnesses of various kinds. The banked fragment, frozen and preserved from deterioration, can literally remake a crucial part of the account holder's body: the blood system."[20]

When combined with the stardust of personalized medicine, the magic of the stem cell technologies casts a powerful charm. But how genuine are these promises?

Of the nine hundred thousand cord blood units in private banks worldwide, only about one hundred had been used for autologous transplants as of 2010.[21] Advocates of private banking might retort that this is only to be expected: the technology is new, and there's a time-lag factor. But, in fact, the technology isn't all that new: it's about twenty-five years old. Since the first successful cord blood transplant for the child with Fanconi's anemia in 1988, about 14,000 unrelated cord blood transplants from public banks have been performed for patients with leukemias and bone-marrow disorders.

Given the three-to-one preponderance of units banked for personal use over publicly banked units, that's odd—but it's not so odd when you take into account the clinical superiority of allogeneic over autologous blood, for example, when acute lymphoblastic leukemia is present even at the fetal stage.[22] Additionally, it's thought that the graft versus host reaction triggered by a transplant from someone else also confers an immunological benefit.[23] "Therefore, the use of privately stored autologous cord blood transplants may be associated with a higher risk of relapse."[24]

New expansion techniques and multiunit drafts from two or more donors[25] mean that while twenty years ago it was mostly children who received cord blood transplants—because of the low volume of cells available—the majority of cord blood transplants are now performed on adults.[26] Rates of disease-free survival for leukemias and lymphomas treated with a single dose of cord blood—the quantity that private storage would normally provide, though some firms now bank two samples—vary between 23 and 50 percent.[27] That's encouraging but still a considerable distance from the complete cure of such blood disorders by cord blood transplants in 100 percent of cases, an outcome some private banks imply is possible.

For non-blood-related conditions, there is still less evidence of benefit in humans from banking blood privately. One individual case report in 1999 detailed success in treating a girl with a highly malignant brain tumor (neuroblastoma), using blood that had been taken at her birth because her older brother had leukemia.[28] More controversially, several children with cerebral palsy have been reported cured by infusions of their own cord blood banked at birth. But it may not just be coincidental that they all had the same condition.[29] The harsh irony is that their original brain injury might have been *caused* by too-early clamping of the umbilical cord in order to bank the blood, rather than letting full placental transfusion take place as nature intended. We'll return to this crucial issue in the next section.

At the time of writing, clinical trials involving cord blood are going on for children with cerebral palsy, congenital deafness, and traumatic brain injury.[30] In preliminary results from another study, eight children with type 1 diabetes treated with autologous cord blood stored at their birth were found to have lower insulin requirements.[31] But as a comprehensive review article concluded, "However, even with these ongoing investigations, the scientific evidence clearly supports public cord blood donation [over private banking] due to the likelihood of clinical need, potential graft vs. leukemia effect, concern over latent disease in the cord blood unit, and quality of autologous cord blood units."[32]

What about publicly banked cord blood? Most of the current benefit is limited to blood-related diseases, including not only hematological cancers but also thalassemia and sickle cell disease, sometimes through donations from siblings.[33] Given the prevalence of those conditions in some developing countries, publicly banked cord blood could be very valuable, although it's private banking that's attracted the most attention.[34] It's still early days: only thirteen cases of thalassemia have been treated so far in India. However, the survival rate was 83 percent—although all the patients had additional bone-marrow transplants from siblings, undermining the appeal of cord blood as a less invasive procedure than bone-marrow transplant and making it unclear whether cord blood on its own would have worked.

Allogeneic cord blood has also been used to treat inherited metabolic disorders in children, with a one-year overall survival rate of 71.8 percent.[35] In 2010, doctors at Great Ormond Street Hospital in London successfully used an allogeneic cord blood transplant to treat an eleven-month-old "bubble boy," Imtiyaz Ahmed, who had severe combined immunodeficiency syndrome.[36] The boy's older brother Mirza had already died of the condition: his body didn't make enough white blood cells to fight off routine infections. Because the condition is genetically linked, using the child's own cord blood was out of the question: it had to be publicly banked blood. Mrs. Ahmed said afterward, "I'm very grateful to the mother who donated her baby's cord so my child has the chance to be the normal fun-loving child he deserves to be."

However, we're still a very long way from anything like seventy diseases being treatable with cord blood, even of the allogeneic variety. In mouse and rat models, it's true, human cord blood cells injected into the tail vein have been found to migrate to myocardial tissue affected by a heart attack and to reduce the size of the infarct.[37] In another study by Jun Tan and colleagues,[38] mice with a condition like Alzheimer's disease showed reduced brain plaques after low-dose infusions of human cord blood. But no such study has yet been performed in humans.

To sum up the medical evidence, there are genuine existing benefits from publicly banked blood and possibly more to come, particularly outside the original field of blood cancers.[39] For example, there has been some recent progress in treating type 1 diabetes in adults with cord blood–derived stem cells in a very small study of fifteen patients and three controls.[40] However, neither autologous nor allogeneic transplants can treat, still less cure, the huge range of conditions claimed by some private banks. Apart from blood cancers and some metabolic conditions, "other uses for CB remain speculative and it is premature to speculate whether non-haemopoietic stem cells are present in cord blood in sufficient numbers for use against degenerative conditions, as is currently postulated by some commercial organisations."[41] The low chance (one in 2,700, according to the American College of Obstetricians and Gynecologists guidelines) that a child will actually need the personal banked sample—because the range of conditions for which it could be used is comparatively narrow—radically diminishes the value of privately banked blood.

As an editorial in the American College of Obstetricians and Gynecologists journal states, "We argue for public umbilical cord blood banking as a matter of good public health and economic sense."[42] The public-health arguments, which lead the authors to recommend that obstetricians should discourage patients from storing cord blood for their own child's speculative future use, have to do with the way in which private banks indirectly poach on public ones by reducing the availability of donations from which all can benefit. By blocking researchers from access to those samples, private banks indirectly threaten the very research on which their marketing plays so heavily.[43]

Economic arguments were also very telling in another study, which found that "Private cord blood banking is not cost-effective because it costs an additional $1,374,246.00 per life-year gained."[44] Using generous estimates of a 0.04 percent chance of requiring an autologous stem cell transplant and a 0.07 percent chance of a sibling requiring an allogeneic stem cell transplant, and calculating on the basis of the cheapest price quoted by any private bank ($3,620), the researchers found that it costs over a million dollars on average to buy an additional year of life by banking cord blood privately. The usual standard for a cost-effective intervention is between $50,000 and $100,000. The price of private banking would have to fall about 93 percent, to $262, before it became cost effective.

Not a single one of the ninety-three hematopoietic transplant pediatricians who replied to a survey would recommend private cord blood banking for a newborn of northern European descent with one healthy sibling.[45] The situation is slightly different for minority ethnicities, because of the greater difficulty of finding a tissue match. Yet even there, only one in ten of these experts favored private banking.

Some might argue that whatever the doctors think, when it comes to saving your child's life, no price you can afford is too high. That wouldn't necessarily be an irrational view: it's based on Pascal's Wager.[46] No matter how low the probability of God's existence, the seventeenth-century mathematician and philosopher argued, it's dwarfed by the magnitude of the possible loss to the nonbeliever of eternal life. Any probability of such a loss is intolerable, if greater than zero. "Do not hesitate, then," counsels Pascal: believe in God. Might not the same be true of your child's possible death? Who would hesitate to spend every last cent to save their child?

The answer is that you would rightly hesitate if it might do the child more harm than good. To assess that surprising and shocking possibility, we need to look at the converse side of the evidence base: possible harms from cord blood banking. And to do that, we need to examine in greater detail how blood is obtained by private and public banks.

THE EVIDENCE BASE: HARMS

"The collection immediately after the birth is totally painless for mother and baby and does not present any risk." "The collection of these precious stem cells is totally safe and harmless to both mother and newborn." These typical extracts from the websites of private cord blood banks (Cryo-care and Cryo-genesis) are echoed in remarks from some experts, such as an editorial in the *Canadian Medical Association Journal* accepting that cord blood collection is risk free.[47] Yet there's actually a great deal of controversy about whether that's accurate, or whether the collection procedures actually *impose* risk on the baby and possibly also on the mother.

Particularly with private cord blood banks, cord blood is collected during the final stage of childbirth, between the delivery of the baby and that of the placenta. It's only accurate to say "the collection immediately after the birth" if you concentrate only on the delivery of the baby. But the final stage is medically critical.

> While to the exhausted labouring woman this stage may be an afterthought, it is a crucial time for fetal-to-neonatal transition. Major changes in anatomy and physiology occur in both mother and baby. It has also been described as "potentially the most hazardous time of childbirth," largely due to the risk of postpartum haemorrhage (PPH) on placental separation.[48]

Postpartum hemorrhage is the main cause of maternal death, accounting for nearly a quarter of fatalities worldwide.[49] The first breaths, fetal adaptation to the outside world, and safe expulsion of the placenta are all complex processes. If delivery room staff are distracted from their main tasks by pressure to fulfill a contract that parents have made with the cord bank, it could well be dangerous.[50]

But that's not all. Cord blood is often presented as mere waste that would otherwise be discarded—so that on the principle of "waste not, want not," it could and should be harvested to meet growing clinical demand.[51] The picture conjured up is of discarded cords and placentas being "recycled" rather than thrown on the clinical waste heap. Actually, however, what happens during the third stage is that the cord is clamped, stopping the natural continued flow of blood from the placenta to the baby. The cord may be clamped early (within the first minute or so) or late (after more blood has been allowed to flow to the baby). If cord blood is being taken, that will be done after the clamp is at-tached. Timing of the clamp is one variable; the second variable is whether the cord blood is extracted while the placenta is still attached to the uterus (*in utero*) or when the placenta has been delivered (*ex utero*).[52]

So the issue is whether early clamping combined with *in utero* collection—the most common practice in private cord blood banks—interrupts the natu-ral flow of cord blood, a crucial substance, to the baby. As the obstetrician David Hutchon wryly puts it:

Fisk and Atun point out that "demand for stem cells from cord blood is greater than supply." This is true. The demand is from the baby but the sup-ply is usually artificially manipulated by the cord clamp. This may have very serious consequences for the newborn baby. The solution in this situa-tion is simple. If the baby is allowed to receive as much blood from the pla-centa as it requires then the supply will always be sufficient.[53]

Although clamping within a minute or less of delivery has become widely accepted as part of active management of the final stage of labor, there's in-creasing concern among obstetricians that it deprives the baby of up to 60 per-cent of red blood cells and 30 percent of blood volume.[54] For a full-term in-fant, allowing full placental flow gives the baby an additional eighty to one hundred milliliters of blood.[55] A series of clinical trials found that delayed clamping could reduce the risk of anemia, chronic lung disease, brain hem-orrhage, sepsis, and eye disease in later life.[56] While early clamping slams the brakes on hard and fast, "leaving the umbilical cord unclamped allows a pe-riod of transition between the fetal and adult circulations . . . thereby easing the newborn into extrauterine life."[57]

An infant who has received the full iron ration from placental blood flow is less likely to require resuscitation at birth or to develop infant respiratory distress. Iron-deficient infants can take a long time to recover, continuing to perform less well on developmental tests than those with sufficient iron re-serves.[58] A systematic review of eight trials from both developed and develop-ing countries found that anemia persists until two or three months of age in infants whose cords were clamped early, and that early clamping contributed to a 15 percent greater risk of developing the condition. Where mothers are anemic themselves, as in much of the Third World, this risk is particularly severe.[59] A randomized clinical trial conducted in Mexico found that the difference in hemoglobin counts between early-clamped and late-clamped babies still persisted at six months.[60]

"All the evidence indicates that there is *harm to the newborn baby by clamping the cord immediately.*"[61] This evidence, which has mounted since early clamping became standard clinical practice, has driven professional bodies, including the Royal College of Obstetricians and Gynaecologists, to issue a strong recommendation against early clamping.[62] Other health bodies, including the American College of Obstetricians and Gynecologists, the International Federation of Obstetrics and Gynecology, and the World Health Organization, no longer recommend immediate cord clamping, for the same reasons.

So why did they ever recommend it? The RCOG admits that "immediate cord clamping became routine practice without rigorous evaluation."[63] Early clamping developed into normal practice as part of a package of "active" versus "expectant" management of childbirth.[64] While some aspects of active management do reduce maternal risk—including administration of uterotonic drugs such as oxytocin to lower the chances of maternal hemorrhage—early cord clamping carries no such benefits. A systematic review in 2001 found that while prophylactic oxytocin did reduce the risk of postpartum hemorrhage, early clamping did not improve the risk of maternal hemorrhage and so could be abandoned if it risks harming the baby.[65] (A previous study, not included in that review, had actually found that early clamping substantially *raised* the risk of maternal hemorrhage.)[66]

In fact, early clamping was known to cause harm to the baby as far back as the time of Charles Darwin's doctor grandfather Erasmus, who wrote: "Another thing very injurious to the child, is the tying and cutting of the navel string too soon; which should always be left till the child has not only repeatedly breathed but till all pulsation in the cord ceases. As otherwise the child is much weaker than it ought to be, a portion of the blood being left in the placenta, which ought to have been in the child."[67]

To sum up: set against the speculative benefits of privately banked cord blood, these definite harms from early clamping and cord blood collection tilt the risk-benefit equation decisively *against* individually banked cord blood. The risks can be lessened by practicing late clamping and *ex utero* blood collection—the procedure recommended by the RCOG guidelines.[68] However, in general private banks use the *in utero* method, while public banks tend towards *ex utero* (with some exceptions). The public London Cord Blood Bank, for example, harvests cord blood immediately after delivery of the placenta, using the *ex utero* method by suspending the placenta and attached cord, allowing the blood to drain by force of gravity.[69]

Why do private banks often use the riskier *in utero* method? Particularly when parents are paying for their "money's worth," or if two samples are taken, blood volume matters. A belief seems to persist that collecting the blood while the cord is still pulsing, still attached to the uterine wall, produces more blood for the parents to bank. In a randomized clinical trial conducted by the private blood bank Eurocord, significantly more blood was indeed collected while the placenta was still attached to the uterine wall.[70] However, other studies show no difference between the two methods in cord blood volume collected.[71]

In public banks, there should be less pressure to maximize the donation, since cord blood is immunologically naive. It doesn't react strongly to tissue from another body, making pooled donations effective, although there still needs to be some tissue matching. So delayed clamping and the *ex utero* method would work perfectly well for public banks, you'd think, and indeed that has traditionally been true. However, there is also some evidence that some public banks are increasingly moving over to *in utero* collection.[72]

To understand one possible incentive for them to do so, we need to look at the way cord blood has become a commodity in the international "bioeconomy." That in turn raises the troubling question for We Medicine of whether women are being encouraged to make an altruistic gift of something that is then commodified into a form of "biocapital." Is their concern for others being exploited for commercial gain?

GIFT, CORD BLOOD, AND BIOCAPITAL

In an astounding and comprehensive study, the British sociologists Nik Brown, Laura Machin, and Danae McLeod have found that altruistically donated cord blood units in public banks now trade internationally at very high prices: between $23,000 and $31,000 a unit.[73] At the end of 2008, this global commerce was worth over thirty million dollars—and the "lead players" are public banks in North America, Europe, and Asia. Before you dismiss their hypocrisy out of hand—as I must admit I was tempted to do, when I first read this study—here's why they do it.

What's driving these public banks isn't making a profit or even simply covering storage cost—which is usually less than one-tenth the export price of a cord blood unit. The real driver is the rarity value of immunologically typed blood for ethnic minority populations. For a long time, those populations have been underrepresented in public bone-marrow registries[74] and indeed in tissue donation more generally. That imbalance drives some commentators to argue that a market in organs is actually the fairest mechanism to serve African Americans.[75] The probability of finding a suitable match in the U.S. national bone-marrow registry is only 27 percent for them, compared to 75 percent for whites.[76]

Many public cord blood banks have tried to compensate for a similar inequity by ensuring high representation of ethnic minority groups.[77] This strategy now gives them a valuable trading counter with other public banks and enables them to afford to buy tissue-matched blood not in their own collections, as need arises—particularly crucial when cord blood transplants are performed more and more often in adults, who need multiple units. Over 40 percent of all cord blood units used are traded across borders. Very few national banks are diverse enough in tissue types to meet domestic demand completely: only the more ethnically homogeneous nations (Japan, China, and Korea).

However, the temptation, even for public cord blood banks, is to concentrate not on supplying their own nationals but on building up an export business. Beset by competition from private banks and cutbacks in government funding, public banks feel under pressure to make themselves economically self-sufficient. The binary "Me and We" polarization of private and public banks looks increasingly blurry, as costs to public banks are countered by trading units commercially to other public banks. Those banks that do best are the ones that have the most valuable "corporeal currency": ethnic diversity in cord blood samples is a trade advantage. "So while cord blood may be the raw material of the market, the actual asset is race itself."[78]

That sounds familiar—much as BiDil tried to capitalize on racial identity. But actually it's a different phenomenon. Whereas BiDil was marketed on the strength of a racial "brand," it wasn't specifically developed for African Americans and was in fact just as effective (or no more effective, compared to the component generics) in whites. But in the "immunitary bioeconomy" of cord blood, tissue matches do genuinely matter—even allowing for the comparative plasticity of cord blood—and they differ widely by ethnicity. Many public cord blood banks do specifically develop their "product" for particular minorities.

The United Kingdom's NHS Blood and Transplant Bank, which recruits potential donors from five London hospitals with high ethnic minority birth rates, has been very successful at incorporating a wide ethnic range—allowing it to export more widely than most other national banks. It has the second highest percentage in the world of rare immune types (over 40 percent). However, successful exporters aren't necessarily good at satisfying their own domestic demand: hence the United Kingdom's overall trade deficit of about 70 to 80 percent. Banks in Germany, Belgium, Australia, and the United States all now have a significant trade surplus, while Canada actually has a 100 percent deficit.

Brown, Machin, and McLeod point out that while private cord blood banks can be accused of reinforcing *class* privilege—only those who can afford them need apply—the trade among public banks reinforces *national* privilege. All the countries that take part in the trade are from the developed world. As far as any sense of We-ness with the poorer countries goes, they can be accused of retreating into Fortress First World. Their altruism is limited to their own nationals and to citizens of those other countries that can afford to trade with them. Solidarity, community, and all those other We values have little to do with it. "The international trade in cord blood is not necessarily a freely given expression of common community. It is instead a form of protection for the trade's participants from the vulnerabilities of being dependent on an import market in premium goods."[79]

So are these hard-headed public banks cynically exploiting the altruism of those who freely donate the resource that produces their tradable wealth—the women who donate cord blood as an additional part of giving birth? Rarely do these women know that the blood they donate is likely to become a tradable commodity.[80] Or is the banks' strategy just a sensible recognition of commercial reality? Perhaps it's even a virtuous circle: targeting a representative range of ethnic minorities gives the banks a marketable "product," which in turn assures their continued existence and ability to serve patients who couldn't afford private banks even if they wanted to. Couldn't this practice count as a form of We Medicine?

In interviews carried out by the sociologist Helen Busby with women who donated cord blood to a UK public bank, it's clear that the mothers were indeed responding directly to the message that they had been given: "your birth day gift, helping to save a life." They were asked to bestow a "double gift": life for their own newborn and also life for another child who could be saved by their altruism. It's that other baby like their own who's foremost in their minds: their narratives make it clear that they were reaching out empathetically to another object of their maternal concern, in what Busby calls "a commitment to mutuality." Where there's a particular emphasis on recruiting donors from minority ethnic backgrounds, as there was in this bank, these women may also be expressing racial solidarity.

The slogan "your birth day gift" doesn't immediately translate in most people's minds as: "Please give us, free of charge, a unit of cord blood that we can trade on the international markets for $25,000." Whether women would have donated if they were approached that way is at best uncertain: the possibility of the unit being traded wasn't mentioned in the leaflet they received. While the mothers certainly recognized the value of cord blood, that value wasn't expressed in terms of dollars and cents or pounds and pence. Rather, the value to them lay in doing something selfless.

In the words of one woman: "It didn't involve a lot from us, because it was something that was valuable for us in terms of, you know, we got to feel like we were doing something useful and helpful and the possibility that maybe if we needed it [in] the future we might be able to access it. And . . . something that would go to waste would be helpful for other people."[81] Here, the respondent is echoing what she was told by the midwife—but what she was told isn't true. As we've already seen, cord blood isn't just clinical waste. Rather, it has to be taken deliberately in an additional procedure. Arguably, the mother was also told a half-truth about whether she would be helping another child directly. If trading blood units commercially decreases the likelihood that her child's own blood will remain in the public bank for her possible future use, as it almost certainly does, then she was also being misled in that respect. Even in the United Kingdom, where the law of informed consent is laxer than in the United States, the cardinal principle has long been that patients need not be told all relevant facts, but they must not be misinformed.[82]

It's true that the trade in blood units helps keep public banks functioning and enables other children to be helped. But the empathetic reality matters, not least because the clinic obviously thinks it does. It's certainly prominent in the heartstrings-tugging presentation given by the midwife who talks to the women about donating:

> The recipients of transplants, often children with acute leukaemias or severe blood disorders, have a central place in this narrative, and those who died waiting for a transplant are also referred to. . . . The narrative that emerges is a powerful one that merges together the missing futures for the children who did not reach adulthood, the futures regained for children who were recipients of successful transplants and the future needs of unknown children and adults. . . . The headlines of patient leaflets, press re-

leases and the midwife's presentation all refer to the double "gift of life" of a newborn who might save the life of a sick child. This is echoed in the following phrase on the cover of the information leaflet for women invited to donate: "Your Birth Day gift . . . Helping to save a life."[83]

The global bioeconomy of cord blood, in which both private and public banks take part, relies on a substance that is either donated for free or that the parents have actually paid to store. It transforms this substance into a very valuable commodity, using the language of "waste" in a manner that disguises that value, and playing on the language of "gift" in a way that ignores the profits made by everyone except the original donor.

Even more value can be added by patents of the sort filed by the Biocyte Corporation as U.S. patent number 5,004,681, for the cryopreservation of neonatal and fetal blood. Another firm, PharmaStem Therapeutics, later acquired the rights to the Biocyte patent and took out international patents in Europe and Japan. It then defended that patent vigorously—to the extent of sending letters to 25,000 physicians, informing them that they would be infringing on its patent if they collected cord blood on behalf of rival private banks—before it was taken to court for interference in the doctor-patient relationship.[84] The judges found against PharmaStem and in favor of the plaintiff, Susan Christopher, whose obstetrician had refused to collect cord blood as she had asked, because he feared that he would be sued by PharmaStem for collecting on behalf of a rival.

True, patents may represent the input of additional skill and inventiveness by the firms involved, but the bottom line is that none of this vast bioeconomy would exist if women weren't donating the cord blood freely or even paying for the privilege. What the claim that cord blood is clinical waste does is to mask the mother's rights in the cord blood, making it appear to be something abandoned, open to the private bank to process and store for a considerable fee. Here we should be reminded of the similar tactics used to claim property rights in so-called discarded tissue by commercial interests in the *Moore* case. There the claim that a three-billion-dollar cell line was mere waste was implicitly accepted by the California Supreme Court in denying the donor (Moore) any property rights in the line developed from his tissue. That case has gone on to create endless controversy, but a similar possibility in cord blood banking seems to pass largely unnoticed.

One iconoclastic way of seeing this upside-down phenomenon is to view it as the transfer of surplus value from mothers to both private and public cord blood banks, all through the mechanism of altruistic donation. Marx thought that the driving motor of capitalism was the seizure by employers of the difference between the value put in by the workers and the final value of the sold product. But at least the commercial gains of nineteenth-century mill owners weren't made off the back of workers' good will. Although the working-class novelist Robert Tressell called his characters "ragged-trousered philanthropists" because they were effectively subsidizing their employer as their wages were cut time and time again, he meant it ironically. No one expected workers to donate their labor out of altruism, just hard necessity.

But in the case of cord blood banking, women's commitment to We Medicine values such as altruism and empathy is being used to create a form of capital held and traded by private and public entities alike. Catherine Waldby and Melinda Cooper analyze the cord blood phenomenon as a form of "regenerative labor," in which women are encouraged more generally to give their tissue—construed as surplus or waste material but with generative powers that shouldn't be withheld from others. We've seen that the international trade in cord blood is confined to the wealthy countries, but in the Third World, Waldby and Cooper note, other forms of tissue such as human eggs or "services" like surrogacy are obtained for the global reproductive tourism trade through payment of a fee well below the profit made. Either way, they think, women's bodily productivity is being harnessed to create biocapital. "In each case, female bodily productivity is mobilized to support bioeconomic research, yet the economic value involved in these relations is largely unacknowledged."[85]

Why isn't the injustice of this more widely noticed? The basic answer is that very little attention has been paid to the mother's contribution. Although a Marxist analysis of exploitation provides the key concept of seizure of surplus value, Marx himself never applied it to women's reproductive activities, which he classified primarily as natural and outside the market realm. It has taken feminist analysts to put that right, demonstrating that in biotechnologies such as egg extraction women put in extensive work and take risks to produce a product that is not simply "natural."[86] But few feminists have extended that analysis to cord blood harvesting, and even they have sometimes erred in supposing that cord blood belongs to the baby rather than to the mother.[87] Many respected property scholars who have considered cord blood have made the same mistake, assuming that cord blood belongs to the baby, on whose behalf it is banked by the parents.[88]

Just on the simple physiological basis that cord blood is taken from the mother's side of the clamp, flowing from her placenta (whether *in* or *ex utero*), that seems wrong. In the United States, an Institute of Medicine of the National Academies report, setting out proposals for the creation of a national hematopoietic stem cell register and national cord blood bank, took the view that the blood belongs to the mother.[89] In the United Kingdom, the Royal College of Obstetricians and Gynaecologists received similar legal advice during the evidence-based consultations that produced its two Scientific Advisory Committee Reports on the practice of umbilical cord blood banking.[90] That advice held that in the common law of the English-speaking countries, if cord blood is anyone's property, it is the mother's.[91]

The legal opinion obtained by the RCOG decisively rejected the assumption that cord blood belongs to the baby by virtue of genetic or immunological identity.[92] It seems clear that Stephen Munzer, one of the foremost scholars in this area, does rely on genetic identity as the basis for ownership rights, as when he claims that both parents have a genetic share in the baby's genome and therefore should share an equal right to control the disposition and management of the newborn's cord blood.[93] But, as the RCOG report noted, genetic identity is rarely if ever the basis of property rights. The cord blood, the report declared, should be regarded as a gift from the mother to the child, not as the child's property by right.

Normally, in the absence of clamping and harvesting of cord blood, the infant would receive all the blood supplied through the conduit of the cord from the mother. The mother is the donor of the blood and the infant the recipient, in the usual case. Where cord blood is taken, a portion of that blood is donated by the mother to the public or private cord blood bank rather than to the infant. It is donated for the infant's benefit, in private directed banking, but it only "belongs" to the infant because the mother has transferred her entitlements in it.

Rightfully, mothers who use a private cord blood bank empower the bank to act as a sort of steward but don't necessarily surrender all property rights. Whether a mother would have a legal claim if a private cord blood bank went bankrupt or lost what is actually her stored property is a controversy waiting to happen. In the *Yearworth* case, examined in chapter 2, an English court found in favor of several men whose stored semen samples were negligently lost by a hospital.

In some cases, courts have held that researchers or professional medical bodies should be the unequivocal owners of donated tissue, on the basis that they have put work and professional skill into processing them into cell lines or other more complicated forms of tissue.[94] However, in the case of umbilical

cord blood, private banks don't generally undertake the labor of collecting—
that's usually done by delivery room staff, in the prevalent *in utero* method.
Nor do they undertake any further work on the samples. There hasn't been
any transfer of title from the mother through the bank's paying her for the
blood. In fact, of course, the parents pay the private banks for storing the
blood. Since they're paying for a service, the parents, particularly the mother,
might even have a stronger claim than the men in *Yearworth*.

In public banking, separate personnel employed by the bank usually carry
out the collection through the *ex utero* method, giving the public banks a
stronger title. But even there, the blood isn't necessarily processed any further. While it's true that the mother has made an *inter vivos* gift of the blood
to the public bank, which would be final,[95] the issues I've raised about whether
her consent was fully informed might possibly invalidate that gift.

If, as Marx thought, productive labor is distinguished by intentionality and
control, the decision to allow cord blood to be extracted requires both those
qualities. Women must decide in advance that they intend this additional
procedure to be performed and that they will be undergoing it at a time when
they simply want childbirth to be over as quickly as possible, because they
intend to confer a benefit on their baby or another mother's child. Little do
they know that their admirably We-minded altruism is actually creating a
form of biocapital.

In previous chapters, I've perhaps been guilty of simplistically assuming
that We Medicine is always a good thing, although I've tried my best to avoid
the corresponding sin of assuming that Me Medicine is always misguided.
But in cord blood banking, the way in which public banking is intertwined
with profit motives and the use of We language to exploit women's altruism
leave substantial doubt about whether We Medicine is quite as snow-white as
I've assumed. You might think that women do enough giving by giving birth.
How much can they, or anyone, be expected to sacrifice for the collective?

THE FOUR HYPOTHESES

How closely does cord blood banking match each of the four possible hypotheses I've put forth to explain the rise of Me Medicine? Let's take them in order
again: threat, narcissism, corporate interests, and choice.

The private cord blood banks generally avoid the bullying rhetoric of
threat—or at least avoid it openly. They prefer the sunnier language of hope,

particularly in those lengthy and implausible lists of cancers and other diseases that can supposedly already be cured by cord blood stem cells. (Of course, the threat of the child's developing those conditions hides behind the hope that banking the cord blood can preserve the child's health.)

Just as we saw in chapter 1 that some people are beginning to bank their own blood privately, fearing contamination of the public supply in the wake of revelations concerning "mad cow disease," so parents are encouraged to regard a private cord blood account as a form of insurance policy for their child. But there's little evidence that the reason why private banks are outpacing public ones in growth is that parents fear contamination of the public cord blood supply. Perhaps they don't even know that there is any such thing as a public cord blood bank. If they did, they'd have no reason for fear: the blood of others is clinically more effective than autologous blood. If anything poses a clinical threat, you might say, it's the child's own blood: for example, if a genetically based leukemia is already present from the fetal stage.

Threat is very much an issue, however, if the procedure used to extract the blood actually increases risks for the baby and possibly also the mother, if the attention of delivery staff is diverted at a crucial moment. When early clamping is combined with the *in utero* method of cord blood extraction, preventing much-needed cord blood from flowing to the baby, the evidence base clearly suggests that there is a threat of jaundice, anemia, and other consequences. That method is more typically used in private banks: in the clinic where Busby's study was performed, the *in utero* method is said to be used for private banking and the *ex utero* method for public.[96]

As in chapter 3, what looked at first like a plausible explanation for Me Medicine in general, the "*narcissism* epidemic," seems inappropriate here. It's hard to see any connection to the "values of self-admiration and self-expression" in a mother's decision to bank cord blood privately for her infant, still less to give the blood to a public bank. The closest we've seen is the comment from the mother in Helen Busby's study that she is gaining something of value from the chance to contribute to a public bank, but it's a considerable stretch to call that narcissistic.

Bowling alone doesn't really work as an explanation, either: despite the rise of private banks, the United States still numbers far more public cord blood banks than any other country. If we're such a self-centered nation, perniciously anemic in the lifeblood of civic togetherness, how do you explain that? Public banks are a form of biocapital, true, but they're also a variety of *social* capital. Putnam argues that social capital has been in decline over the past thirty years, but proposals for a national cord blood bank and the continued rise in units of publicly banked blood contradict that thesis as applied to cord blood. Even the private banks use the appeal of helping your community and furthering the march of science in their sales pitches, as in these two examples:

> The Community Banking service is a way for you to keep a small amount of your baby's stem cells for your own family, but at the same time support your community and potentially contribute to saving someone else's life in the future.[97]

and

> PacifiCord, in connection with its parent company, HealthBanks Biotech, works closely with researchers at a wide range of well known medical institutions, such as UCLA and Johns Hopkins, who are leaders conducting research on the application of stem cells in medical treatments.[98]

That brings us round to the third possible explanation, *corporate interests and political neoliberalism*. It's evident that the medical professional bodies oppose private cord blood banking, but their consistent and frequent pronouncements don't seem to be carrying the day. Some more powerful force is at work. That force can't be parental demand: private blood banks weren't set up because prospective mothers were lining up to demand the possibility. Rather, as I suggested in chapter 1, this is a prime example of supply looking for demand: of capitalism's ability to create demand where none existed before. We met the same phenomenon in chapter 2, in the form of direct-to-consumer genetic testing: the theme crops up over and again in Me Medicine.

Both retail genetics and private cord blood banking are the direct result of neoliberal policies. The Food and Drug Administration relaxed its prohibition on advertising for any pharmaceutical products or services sold direct to the consumer in 1997, and since then this sort of marketing has increased by leaps and bounds.[99] That, along with leaflets placed in clinics and waiting rooms, is how most purchasers learn about private cord blood banking—not so much in consultation with their doctors, whose professional bodies advise against it.

It's striking that private cord blood banking is galloping fastest in countries that pursue broadly neoliberal economic policies, such as the United

States, the United Kingdom, and India—while it's actually banned in France, with its long-standing opposition to "the Anglo-Saxon model" in economics and politics. The French National Ethics Committee continues to oppose private cord blood banking as contradicting the policy of free, anonymous gift of blood and tissues.[100] In an explicit reversal of the neoliberal orthodoxy "private good, public bad," the committee's most recent opinion lauds the solidarity behind the ideal of public banks but condemns private banks in remarkably forthright language: "Their private, for-profit character too frequently incites them to make advertising claims that are not founded on established scientific facts and that are often actually deceitful."[101]

The French government has rejected attempts by private cord blood banks to operate in France, even when they've offered to contribute a portion of the blood to increasing the public supply. This policy has been put on a statutory footing by a law adopted in the National Assembly on October 18, 2011.[102] No doubt that's a disappointment for the private banks, since France has one of the highest birth rates in Europe. But even without the French, the potential market for private umbilical cord blood banking is huge: in countries wealthy enough to afford it, it would include every prospective mother, her partner, and their own parents. The latter are a major "consumer demographic": the private cord blood bank Smart Cells International reckoned that most of its initial customers were grandparents.[103]

Reflecting the promise of this sizeable market, the level of investment by private blood banks is often very substantial, involving megacorporations—such as the Virgin Health Bank, initially funded at £10 million in cooperation with the biotechnology venture capital firm Merlin Biosciences. And as we saw in the case of the Biocyte Corporation and PharmaStem, there could be lucrative additional returns from patents on methods of collecting, processing, and storing cord blood. In the Susan Christopher case, PharmaStem was willing to go to quite considerable lengths—mailing 25,000 warning letters to physicians about not collecting for rivals, plus underwriting the costs of a court case that they eventually lost—to defend those patent rights. So they must be a major incentive, as they often are for the many biotechnology firms whose principal portfolio strength lies in patents.[104]

All these factors combine to make corporate interests and neoliberal government policies the most effective hypothesis so far for explaining the particular form of Me Medicine known as private cord blood banking. Where the neoliberal thesis needs a bit of refining, however, is in dealing with the issues of exploitation of the "workforce" (mothers who donate blood) and in the surprising anomalies of the global trade in umbilical cord blood by public banks. The latter can be explained as rational behavior in a time of shrinking public budgets and of increasing expectations that not-for-profit bodies will be managed and assessed on the basis of the same indicators as private firms. Ethnic disparity has become a form of "branding" and a moneymaker for countries that have a particularly strong mix of populations represented in their national banks. These developments may seem surprising, but they're perfectly compatible with an analysis centered on the pervasiveness of neoliberal policies.

However, a neoliberal model would probably predict less altruistic maternal behavior in mothers than we actually see. Neoliberals might be surprised that mothers haven't yet made any attempts to assert their property in the cord blood they give to private banks for storage. To some extent, that anomaly arises from the general but mistaken assumption among legal scholars that the blood belongs to the baby, but the doctors' professional bodies haven't taken that view: they view the cord blood as belonging by rights to the mother, yet no private blood bank contract is based on that position. Analysis in terms of neoliberalism needs to be supplemented by a feminist approach concentrating on the question of why women's property in the body often goes unrecognized, why "the lady vanishes," and of the particular kind of exploitation that results, particularly when one-way altruism itself becomes a form of exploitation.[105]

There just remains the hypothesis of *choice and autonomy*, which I find largely unconvincing in this case. It certainly wasn't the name of the game for Susan Christopher: her choice of using another cord blood bank, and indeed of banking the blood at all, was blocked by her obstetrician's fear of Pharma-Stem. More broadly, the private banks' advertising literature plays on parental duty rather than consumer choice, on parents' natural desire to do their utmost for their baby. Phrases such as "a once-in-a-lifetime opportunity" or "a miracle of nature that is only available once in a lifetime" actually convey the opposite message: that conscientious parents *must* bank the blood now, that medically and morally they have no other choice. And who would say no to that?

5

ENHANCEMENT TECHNOLOGIES
Feeling More Like Myself

If you could cure what I feel is a very serious disease—stupidity—it would be a great thing for people who are otherwise going to be born seriously disadvantaged.
—James Watson

THE ENHANCEMENT DEBATE EPITOMIZES ME MEDICINE AT ITS most controversial. Speculative technologies in cognitive and physical enhancement are premised on the idea that I have a duty to create the best Me I can possibly be. Some of their proponents also argue that we have an obligation to produce the best children we can.[1] While banking your baby's umbilical cord blood is promoted—albeit misleadingly—as a way to guard against future illness, those who argue for creating the best possible children are concerned not just with negatively preventing disease but with positively engineering the ideal child.

Even some opponents of enhancement technologies rely on the language of individuality, countering that the child's autonomy is threatened in the process of enhancement, with her "Me-ness" fatally undermined. "Eugenic interventions aiming at enhancement reduce ethical freedom insofar as they tie down the person concerned to rejected but irreversible intentions of third parties, barring him from the spontaneous self-perception of being the undivided author of his own life."[2] It's also appropriate to critique the enhancement technologies from a We standpoint, however, as I'll do later in this chapter.

"Enhancement" has been defined as "a deliberate intervention, applying biomedical science, which aims to improve an existing capacity that most or all normal human beings typically have, or to create a new capacity, by acting directly on the body or brain."[3] Alternatively, it's been said to be "any medication designed to improve *individual* human performance and brought about by science-based or technology-based interventions"—making the link to Me Medicine plain.[4]

These technologies include human growth hormones, specialized prosthetic devices, neurocognitive stimulation techniques, "genetic doping" in sport, drugs to enhance mental functioning, and genetic manipulation—either of your own body or, more controversially, germline genetic modification, which will affect your descendants as well. Where therapy ends and enhancement begins is also a controversial issue, with the British sociologist Nikolas Rose warning that "The old lines between correction, treatment and enhancement can no longer be sustained."[5]

But is there really a paradigm shift here? As with other areas of Me Medicine, indeed even more so, a reality check is in order. One study of 142 newspaper articles written between 2008 and 2010[6] found that highly optimistic media reports about neurocognitive enhancement were frequently untroubled by anything so crass as scientific evidence. While 94 percent of the reports portrayed neuroenhancement as already being common medical practice, under half that number (44 percent) actually backed up their claims by citing any authors or journals from the scientific literature. It's probably no coincidence that a whopping 95 percent of the media articles mentioned at least one possible *benefit* of using prescription drugs for neuroenhancement but only 58 percent mentioned any *risks* or side effects. What proof do we have that the wonders of enhancement technologies are anything but media hype?

THE EVIDENCE BASE

I've begun every chapter so far with an account of the state of the art for a particular Me Medicine technology. With enhancement technologies, the task is very daunting indeed—which, on a charitable reading, might explain why so few commentators attempt it. (That applies not just to journalists but also to academics.) Yet it's particularly important to try to get the science right here, because both proponents and opponents of enhancement often get it wrong: they share a propensity to exaggerate.

Even Allen Buchanan, a careful moderate who describes himself as neither proenhancement nor antienhancement—merely "anti-anti-enhancement," dedicated to deconstructing weaknesses in some of the more celebrated antienhancement arguments—falls into this trap when he asserts that the science behind enhancement is so advanced as to be unstoppable. "We are faced with a complex but undeniable fact: something momentous is happening on an increasingly large scale, there is every reason to believe it will continue . . . and there is no realistic prospect of stopping the development in its tracks."[7] Like Francis Collins, with his insistence that "You have to be ready to embrace this new world" of personal genomics,[8] Buchanan claims that resistance is useless.

I'm reminded of the end of Stephen Vincent Benet's *John Brown's Body*, when readers are warned not to judge whether the Union or the Confederacy fought the just fight—only to accept the industrial future symbolized by the North as an inevitable force: "Say not with them, 'It is a deadly magic, and accursed,' nor 'It is blest,' but only 'It is here.' "[9] We saw that this wasn't true of personalized genetic testing. Is it any more true of enhancement? Buchanan doesn't really even try to prove that it is, by systematically reviewing the evidence—even though he criticizes antienhancement writers for making sweeping claims without backing them up with empirical facts.[10] In this crucial omission, he's typical of both sides in the enhancement debate.

With a few exceptions,[11] writers on neurocognitive enhancement generally start from the ethics rather than the science, often leaving the science out altogether. In relation to genetic engineering, both sides are particularly prone to overstate the extent to which we can actually intervene to change the phenome through altering the genome, leaving out the role of epigenetics (as examined in chapter 3). Crucially, and to my mind unethically, they also underestimate the barriers posed by the realities of reproductive medicine, especially the burdens that the genetic engineering of future generations would place on women. I'll return to this point later.

Before we can begin to lay out the evidence for and against the success of enhancement technologies, we need to decide what we're seeking evidence *for*. This is a tricky endeavor: what counts as success is less straightforward than in previous chapters, because the criterion isn't cure for or prevention of disease. Cosmetic surgery, for example, counts in my mind as an enhancement technology, although except in cases such as breast implants after mastectomy, it's not normally linked to any disease. But it's clearly Me Medicine, as one woman testified: "I'd rather spend my money on Botox and a procedure here and there than [on] something that is not a part of me. All we have in life is ourselves, and what we can put out there every day for the world to see. The world is not going to see my great record collection or the stuff I have at home. They're going to see me. And Me is all I got."[12]

Yet in the recent case of the use of industrial-grade silicone in breast implants made by the French firm PIP (Poly Implant Prothèse), it was We Medicine, in the form of the UK National Health Service (NHS), that was called on to pick up the pieces. Although 95 percent of the operations had been performed privately, the private clinics refused to remove the implants unless patients paid for another procedure, leaving the NHS under fire when it agreed to cover removal but not implantation of new prostheses.[13]

Evaluating the evidence base for enhancement technologies is additionally complicated because the variety of those technologies is considerable and confusing. In this chapter, I'll concentrate on two principal areas: *neurotechnologies* and *genetic engineering*. Even those are quite capacious umbrella terms. Under the techniques variously known as gene transfer, genetic engineering, or genetic modification—sometimes also called gene therapy, although some authors think the term "therapy" prejudges the value of the techniques[14]—there's a distinction between *somatic* gene modification, targeting particular subsets of body cells, and *germline* genetic modification, affecting all cells and transmissible to future generations. I'll concentrate on germline genetic modification, which has attracted the most column-inches. Under neurotechnologies, I'll focus primarily on those affecting cognitive function, but I'll also mention neurotransmitters linked to emotions and social interactions.

Neurotechnologies

Neurotechnologies can range from routine pill popping by high school and college students bent on improving their exam performance to the extreme fringe of "transhumanism," which prophesies the resurrection of humans as a new species with vastly more substantial mental powers.[15] In addition to pharmaceuticals, they include such novel technologies as those in the box, some of which have possible uses in both therapy and enhancement.

Some Novel Neurotechnologies

Neurostimulation: These techniques include:

Deep brain stimulation, in which an electrode is placed inside the brain. This technique is currently in use for Parkinson's disease, epilepsy, stroke, and severe obsessive-compulsive disorder. Experimental research envisions uses for obesity, Tourette's syndrome, anorexia, and addictions.

Transcranial magnetic stimulation, a noninvasive procedure involving the application of a magnetic field to induce electrical currents in the brain. Uses at present include treatment for depression as well as for the improvement of cognitive functions such as attention, understanding, and perception. Possible applications include treatment for severe migraine, along with nonmedical applications such as enhancement of mood and improvement of memory or problem solving.

Brain-computer interface: This is a system for measuring and analyzing brain signals, which are then converted into computer-based communication or control of a device. Possible therapeutic uses might include restoring communication with patients who have "locked-in syndrome" or allowing paralyzed patients to control a wheelchair. Nonmedical uses include computer gaming and the euphemistically termed "military applications," such as connecting a missile engineer's brain directly to drone aircraft.[16]

Source: Adopted from Nuffield Council on Bioethics (2012)

Drug-related cognitive enhancement—sometimes dubbed "BoTox for the brain" or "cosmetic neurology"[17]—is more accessible and less speculative than any of these technologies. Indeed, it's already here in the form of pills such as Ritalin (methylphenidate) and Adderall (amphetamine).[18] And it's the subject of an "ethical framework" from the Ethics, Law, and Humanities Committee of the American Academy of Neurology (AAN).[19] These guidelines were intended to address the questions raised by the way in which drugs originally developed for the treatment of attention deficit disorder, Alzheimer's disease, and other neurological conditions are now being requested by healthy patients, to improve their normal functioning. That scenario was explored in the 2011 film *Limitless*, which asked what would happen if "smart drugs" allowed us to use 100 percent of our brains 100 percent of the time. (The results weren't pretty.)

Stimulants such as methylphenidate can be used to improve performance on academic tests or to learn new skills. Cholinesterase inhibitors (e.g., donepezil) can treat normal age-related memory changes and have also improved the performance of commercial pilots in flight simulations. Other professionals, such as surgeons, are also interested in neurocognitive pharmaceutical enhancement.[20] Although these drugs are approved by the FDA, such additional uses would be off-label and controversial. The authors of the AAN report write that "This report and guidance should not be construed either to promote or discourage the prescription of neuroenhancements."[21] Yet the mere fact that such a statement has been issued by a professional body might seem to indicate that there is a reputable evidence base for the benefits of neurocognitive drugs administered to healthy normal individuals, and this helps legitimize the practice. But there's also been a backlash against the AAN's statement on the grounds that it's premature.

One reason why it's premature is that the evidence base is at best sketchy and at worst negative. Under "Neurology: Cognition-based Interventions," the Cochrane Library lists the following relevant "meta-analyses," each an international review of all the studies done on that particular subject. The Cochrane Library is respected as the most authoritative and systematic repository of systematic clinical trials, so it's fair to say that this sums up the evidence base about cognitive enhancement, broadly construed, at the time of writing. Here's their summary of the evidence:

1. A study about cognitive training (not drugs) for both normal people and those with cognitive impairment concluded that firm findings were not available;[22]

2. Research testing the growing public enthusiasm for supplementation with the adrenal steroid hormone DHEA, as a means of retarding aging and age-associated cognitive impairment, demonstrated that there is very little evidence from controlled trials;[23]

3. A review of the use of folic acid supplements to improve cognitive functioning in healthy older people found no evidence of any benefit;[24]

4. Negative but solid evidence shows that hormone replacement therapy does not protect against cognitive decline in older women with normal intellectual ability;[25]

5. Another negative finding reports no evidence that dietary or supplemental omega-3 polyunsaturated fatty acid reduces the risk of cognitive impairment or dementia in healthy older people.[26]

Want to know what does help improve your cognitive abilities? At the risk of sounding smug, as a confirmed gym nut, I have to tell you that what works is working out, not drugs. The only positive finding in the Cochrane database about what improves cognitive functioning in healthy people, specifically older people, was increased physical activity—with the largest effects in a comprehensive review of eleven randomized clinical trials shown on cognitive speed and auditory and visual attention.[27] By contrast, the comprehensive reviews listed above show that there is very little positive evidence for drug enhancement to improve cognitive functioning and, in some cases (e.g., 3, 4, and 5), there are negative findings about substances that had seemed likely contenders.

The other reason why the AAN guidelines could be premature is that medical professionals disagree on the ethics of whether they should prescribe smart drugs. "Isn't it putting the cart before the horse to lay out an ethical framework for prescribing cognitive enhancers before we have agreed whether it is even ethical to use them?"[28] The AAN "ethical framework" dwells on principles rather than the evidence base but assumes that society will probably come to see enhancement drugs as acceptable, because the benefits will be found to outweigh the risks. That's prejudging the issue, argue doctors from the Mayo Clinic. "The guidance assumes that the benefits of using neuroenhancers will prove to outweigh the risks in the absence of any evidence that this is the case. However, the principle of nonmaleficence dictates that the use of these drugs by healthy people should not be condoned before reliable evidence for their short and long term safety and efficacy is at hand."[29]

In fact, it's difficult to see how a gold-standard evidence base, a meta-analysis of several randomized clinical trials, could ever be amassed in the case of drugs that aren't routinely tested for their off-label purposes through clinical trials. With some exceptions, such as an experiment in which commercial pilots on donepezil performed better than pilots on placebo,[30] the clinical trial evidence base for neurocognitive enhancers is thin—although it would certainly be in the interests of pharmaceutical companies to do something about that, backing more clinical trials of smart pills.[31] Whether such trials would meet international research ethics standards, such as the principles of the Helsinki Declaration, is another matter. Some might view inflicting risk on research subjects for enhancement rather than curative purposes as unethical in itself.

In an age of evidence-based medicine,[32] the absence of gold-standard evidence might in itself seem a powerful argument against cosmetic neurology.

Clinical trials aim to establish risks, effectiveness, and side effects from a particular drug, things that can't be discovered systematically through mere anecdotal evidence from users. In trials for treating conventional disease, patients might be willing to undergo more unpleasant side effects and to take greater risks than they would for enhancement purposes. In the world of enhancement, research subjects would need to make more complex calculations of how much harm they're willing to tolerate when the alternative is normal health rather than illness, but at the same time they have less objective information to go by.

What we do know indicates there's no such thing as a free lunch. Modafinil, for example, enhances alertness on some tasks but worsens it on others.[33] The dopamine agonist bromocriptine improves performance on "executive-function" tasks for individuals with lower working memory capacity than average but actually *impedes* it in people with higher memory capacity.[34] Genetically modified mice have better memories but feel more sensitivity to pain.[35]

However, as one neurologist points out, "patients' impressions of the quality of their lives do not always correspond directly to bio-markers and symptoms of disease."[36] That's true even for patients with a definite diagnosis of disease, not just for those who want neuroenhancers on a "recreational" basis. For example, "the cardinal symptoms of Parkinson's disease most responsive to dopamine agonists are not necessarily those that bother patients most. Measures of disease activity may not be the best indicator of the impact of multiple sclerosis on patients."[37] These more subjective assessments of symptoms and their relief can't easily be measured in clinical trials, but they still matter to patients with these diseases.

So who's to say that subjective assessments of mental functioning shouldn't matter for patients who *don't* have a disease? It might seem dismissive to bundle their concerns under the label "cosmetic neurology." Particularly in cases on the nebulous borderline between therapy and enhancement, that label may appear condescending and callous. In one pilot study, emergency room patients who were given propanolol after a traumatic event suffered fewer post-traumatic-stress-disorder symptoms, when compared one month later with a control group.[38] Isn't this a legitimate objective of therapy for trauma victims? Yet the studies also suggest that less disturbing memories can likewise be smoothed away,[39] raising profound questions about where to draw the line before we enter the compulsorily cheery dystopia of *The Truman Show*.

Conversely, drugs aimed at enhancing rather than limiting the way our brains encode and store long-term memory of events and facts may also backfire. Drugs that target the transcription factor CREB (cyclic response element binding protein) might increase memory capacity.[40] So can drugs that increase acetylcholine, the main neurotransmitter regulating memory capacity.[41] But more memory capacity doesn't necessarily produce better cognitive functioning.

Cluttering the mind with every experience that we ever underwent would actually impair our mental functioning. "The limits we have in our capacity to remember only so many facts and events may be necessary for an optimal balance between the storage and retrieval of memory."[42] The famous neurologist A. R. Luria had a patient called Shereshevskii, who could remember vast reams of facts but couldn't process any new information. The only work he was fit to undertake was as a performing "memory man" on the stage, the equivalent of "Mr. Memory" in the classic film *The Thirty-Nine Steps*.

To assume that bigger is always better is no more true of our minds than it is of strawberries, where the giants are usually woolly and bland in taste. But "more is better"—faster mental processing, more memory capacity, longer periods of concentration—does seem to be an assumption frequently indulged in by proponents of cognitive enhancement. This might be all very well when you're buying a new computer, but it doesn't necessarily work for human intelligence. Indeed, the French philosopher Michèle Le Doeuff questions whether intelligence is an endowment of nature or an attribute of our own making through our life experiences. As she puts it, "We don't just receive intelligence, we create it for ourselves."[43]

Genetic Engineering

If intelligence isn't an endowment of nature, many of the usual claims for genetic engineering will also be undermined. Even if it were possible to select for traits such as intelligence and pass them on to succeeding generations, that wouldn't necessarily produce what we commonly regard as intelligence in offspring. In fact, the conventional objects of enhancement, such as intelligence, are notoriously difficult to model in the laboratory.[44] But in any case, we're still a long way from the possibility of producing superintelligent future generations.

It was only in 2009 that scientists were first able to genetically modify primates—New World marmosets in this case—and to show that their offspring retained the altered trait, a foreign gene coding for green fluorescent protein, in at least some of their tissues.[45] While similar experiments had succeeded earlier in mice and rabbits, even with mice the success rate still rarely exceeds 10 percent and is often much lower.[46] Marmosets are one step closer to humans—although the scientists who performed the research were careful to note that they're not all that similar to humans and that the findings are mainly relevant to transgenic primate research for modeling illnesses. Other scientists also welcomed the findings as a way of establishing specialized primate colonies to study human diseases such as cystic fibrosis and Alzheimer's disease, which can't be modeled in mice.[47]

The marmoset research was genuinely innovative: no previous study had shown transmission of foreign DNA to gametes—sperm and egg—as is necessary for the animals to breed and establish colonies. But for any transhumanists itching to see this as proof in principle that we could breed a superior kind of human, there are insurmountable problems about the way in which the offspring were created.

A viral vector was used to inject an enhanced green fluorescent transgene into marmoset embryos. The embryos—some created by normal intercourse, some through IVF—were gestated by females acting as "surrogate" mothers. Four out of five of the resulting transgenic marmosets "expressed" (showed) the fluorescent gene in their tissues as neonates. Of these baby marmosets, only one grew up to produce sperm that was used to create a healthy transgenic infant of his own, demonstrating the power to alter genes across generations.

But that one marmoset dad was produced at a cost that would be completely unacceptable in human females—or at least I hope it would. Of a total of 272 IVF cycles producing 121 fertilized oocytes, only one IVF pregnancy and four pregnancies through natural fertilization were achieved.[48] That one IVF pregnancy required no fewer than 460 eggs. These figures aren't far off those in the case of Dolly the sheep, even though the technique is slightly different: Dolly was produced through reproductive cloning, injecting a somatic cell rather than a sperm into an enucleated egg. In the Dolly case, four hundred eggs were enucleated, resulting in 267 usable eggs to be implanted into surrogate mother ewes—all to produce only one successful result. (That's assuming you can call it successful, given that Dolly aged faster and died younger than a normally conceived sheep.) So transgenerational enhancement would entail a major moral concern about the control of women's reproductive capacity. The little-known secret of germline genetic engineering is that it could require stupendous numbers of surrogate pregnancies to ensure that a child with the desired transgenic mutation is born.

Female New World marmosets reach sexual maturity at one year and can have forty to eighty offspring during their lifetimes, with a gestation period of

144 days. Human females don't have—shall we say?—that much leeway. If an attrition rate similar to that in the marmoset experiments were found to occur in humans, it would have required 120 unsuccessful pregnancies, presumably in 120 women, to get that one success. Furthermore, some of the miscarriages were probably caused by cancer-causing genes activated by the technique, leading other scientists to conclude that it would be "unwarranted and unwise"[49] to apply transgenic techniques to humans.

There are alternatives to which transhumanists might resort, but they're not all that attractive either. For many years, it's been possible to select a fetus with a particular genetic variant through preimplantation genetic testing (PGD). In this technique, couples at known risk of carrying a serious genetic condition—such as cystic fibrosis or Tay-Sachs disease—undergo in-vitro fertilization (IVF) in the hope of producing embryos without the condition. In the case of these two genetically recessive conditions, if two parents are both carriers of the harmful variant of the gene but don't display the condition themselves, on statistical average one of every four embryos would manifest the disease. Two would produce "carriers" like the parents, and one would be completely free of the condition, neither a carrier nor a sufferer. The hope would be to produce at least one embryo in this last category and to replace it in the uterus. In germline genetic engineering, you'd be selecting *for* a particular trait rather than trying to eliminate one, but the PGD technology would be the same. That method, however, would require women to undergo all the rigors of IVF, including ovarian stimulation, with its attendant and occasionally fatal risks.[50]

Noninvasive fetal testing, a recently developed technique only requiring a maternal blood test, wouldn't require IVF and PGD; furthermore, it could be done at a much earlier stage of the pregnancy than other screening techniques such as amniocentesis or chorionic villus sampling. But the technology was never intended for enhancement purposes: rather, it was developed for screening in high-risk families with inheritable monogenic illnesses such as Huntington's disease, sex determination in the case of X-linked diseases such as hemophilia, and routine prenatal checking for Down's syndrome and rhesus factor.[51] In December 2010, it was announced that an entire fetal genome had been screened, from the mother's blood sample, for beta-thalassemia (a serious genetic condition affecting many people of Mediterranean origin).

Given the severity of the disease, many women would be willing to abort a fetus with the adverse allele for beta-thalassemia. But would they be equally willing to abort a fetus that merely carried the unenhanced version of a gene that was the subject of genetic engineering? Would obstetricians? How would abortion laws deal with the issue? It's remarkable how few of these practicalities of *reproductive* medicine feature in some proponents' treatments of germline genetic engineering, even those that are up to date on *genetic* medicine.[52] It sometimes seems as if there's a mental disconnect here—rather like the phenomenon I've called "the lady vanishes," which describes how ethical issues about women's necessary role in providing eggs for the stem cell technologies were overlooked by proponents and opponents alike.[53]

Julian Savulescu recounts in another article how he and his wife, as self-avowedly "very risk-averse" and clearly conscientious expectant parents, decided that they would *not* opt for amniocentesis, which can be used to test the second-trimester fetus for genetic abnormalities such as Down's syndrome or spina bifida.[54] As he describes it, they made that choice once they learned that the procedure was four times more likely to result in spontaneous abortion than in detection of genetic abnormality. I'm not criticizing their decision at all: indeed, I find Savulescu's account of his personal dilemma admirably open and honest. The difficult decision seems entirely reasonable in its own right—it's just at odds with the spirit of using preimplantation genetic diagnosis to produce "the best children we can."

Perhaps that duty is to be limited to those already undergoing IVF. However, in his article on "procreative beneficence," Savulescu doesn't give that impression. The principle of "procreative beneficence" doesn't itself specify any limitation about physical abnormality: it only says that "couples . . . should select the child, of the possible children they could have, who is expected to have the best life, or at least as good a life as the others, based on the relevant, available information." In his conclusion, Savulescu asserts that couples should be provided with information about non-disease-related genes through both PGD and prenatal genetic testing before conception.[55]

Of course, prospective parents could still choose not to act on the information, on the lead-a-horse-to-water model, or even refuse to receive it in the first place. In Savulescu's scheme, I presume, there would still be such a thing as the mother's informed consent to testing for genetic abnormality through an invasive procedure such as amniocentesis, just as there is no coercion to choose the "best" fetus. Presumably, if you deliberately choose not to have all the available information, you can't select one child over another—and again, that's not an unreasonable proposition. But it undermines the duty of "procreative beneficence" quite substantially.

Reproductive decisions are tough; those involving genetic abnormality are tougher still. Where disease is produced not by a single gene or chromosome mutation, such decisions would be even harder to make—and most genetically linked disease is associated with several genes rather than one. As I've already suggested, these choices would perhaps be hardest of all when what's at stake is enhancement rather than prevention of disease. Survey evidence suggests that prospective parents do realize that—and draw back from the brink. Savulescu surmises that once couples have decided to use IVF to eliminate serious disease, they would be more willing to employ it to engineer desirable nonmedical traits But in fact, a survey of one thousand parents attending a genetic counseling clinic showed that only 13 percent would test their IVF embryos for non-disease-related conditions such as intelligence and height, compared to the very large majority who preferred to test for disease-linked genes only.[56]

Somatic gene transfer has a thirty-year history, for much of which it was seen as having "extraordinary potential"[57] and as ethically uncontentious, until high-profile scandals such as the death of eighteen-year-old Jesse Gelsinger during gene "therapy." Although it has been used to treat children with "bubble boy syndrome" (X-SCID, or X-linked severe combined immune deficiency syndrome), five children "cured" by the method developed a rare leukemia linked to the gene transfer, and one subsequently died.[58] Even in the hopeful days for somatic gene transfer, however, germline gene transfer was regarded with some suspicion and indeed remains outlawed in many countries.

That position seems to have altered somewhat recently,[59] with germline gene transfer becoming more "chic," despite the scientific limitations that render it almost entirely hypothetical. The nearest feasible development is ooplasmic transfer, which involves transplanting mitochondria from viable human eggs to affected women. Because mitochondria contain their own genome, which is transferred in the process, this technique can be seen as germline genetic modification.[60] However, it's not undertaken for enhancement purposes but as a treatment for infertility or mitochondrial disease.

Germline genetic engineering is also highly contentious in ethical terms.[61] Not only can it be seen as a permanent invasion of the rights of future generations: the techniques by which it would be achieved could put huge numbers of women at risk of harm or exploitation, as the marmoset experiments suggest. This is what occurred in the case of Hwang Woo Suk, who used 2,200 eggs in his fraudulent stem cell research without most of the scientific community blinking an eye (see chapter 1). Commentators in the extensive enhancement literature likewise leave the moral issue of women's exploitation out of the equation almost entirely. Instead, they concentrate on two sets of ethical arguments, which can be handily summed up as being about Me and We.

Before I move on to these two types of argument, let me briefly sum up the evidence base about enhancement, in the much-touted forms of neurocognitive technologies and genetic engineering. With the exception of a few drugs like Adderall and Ritalin, some of them used off-label by FDA standards, neurocognitive enhancers are not yet in widespread use and not comprehensively trialed. They don't come anywhere near meeting the gold standard of meta-analysis of many randomized clinical trials, even though there might eventually be commercial interests backing such studies. There is professional disquiet about premature guidelines for prescribing before the drugs are thoroughly studied. Successful genetic engineering, in terms of transgenes transmitted to future generations, is limited to one monkey in one study with a huge attrition rate. While there are alternative techniques for genetic engineering such as PGD, that would require IVF to become a routine procedure (and incidentally to improve its current success rates beyond the 30 percent mark for women under thirty-five).[62]

The most succinct summary of the enhancement evidence base is that it's very much like the man upon the stair: "The other night upon the stair / I met a man who wasn't there. / He wasn't there again today." It's up to you, whether or not you concur with the poem's final line: "I really wish he'd go away!"

AUTHENTICITY AND PERSONAL AGENCY: ENHANCED BY ENHANCEMENT OR THREATENED?

If the enhancement technologies ever actually achieved what's already being claimed for them, would they really create "the best Me I can possibly be"? Or would that "better" individual actually be someone else—not "Me" any longer? Even if the transformation weren't so profound as to threaten my core identity, perhaps the fact that I'd relied on drugs to improve my intelligence would mean that I can't now claim the credit for being smarter: it's not through my agency but that of the drugs.[63] After all, athletes who are found to have taken performance-enhancing drugs can lose their medals on similar grounds.

We've already encountered Jürgen Habermas's contention that genetic engineering of future generations would undermine the agency and autonomy of those yet unborn. A related but distinct concern is whether those "enhanced" persons would wonder whether their achievements really belonged to them.[64] Similar considerations about the authorship of one's achievements apply to those who've enhanced their cognitive abilities through drugs or novel neurotechnologies.

Or do they? Presumably Einstein, like any good Swiss, enjoyed *Kaffee* with his *Kuchen*, but it's never been suggested that he should be stripped of the genius label on the grounds of his caffeine consumption. I'm partaking of the same brew as I write these words, but I still consider myself indisputably their author. Isn't caffeine also a neurocognitive enhancer? So what's the difference, except that caffeine is more widely used?

Advocates of enhancement sometimes extend this argument to any social institution whose goal is to improve human productivity, awareness, or intelligence—including education, agriculture, and legal systems.[65] If benefiting from these is permitted, even admired, then what's wrong with using modern neurocognitive technologies to achieve the same ends? One answer is that this is to include so much under the rubric of "enhancement" as to render the term meaningless. Another, directly related to We versus Me, is that these are actually social systems and communal achievements, whereas enhancement is defined very much in individual terms.

Nevertheless, this argument leads into an interesting speculation: that both opponents and proponents of enhancement subscribe to a similar conception of human flourishing—one rooted in bringing out our best, most authentic selves. Erik Parens believes that both sides share the same ideal of being true to yourself but define that authenticity somewhat differently.[66] Yet we need to ask if whether being true to yourself is necessarily the highest value. Could it be construed as a more sophisticated form of narcissism? Why equate "best" and "most authentic" selves? Are we so narcissistic that we feel our true selves are our enhanced ones? The point of enhancement, then, would be making me feel more like myself—my "real" self. As one woman exclaimed after her "extreme makeover" operation, "Oh my God, I finally look like me!"[67]

In his book *Better Than Well*, Carl Elliott likewise explores the link between enhancement and cosmetic surgery, whose devotees often say it makes them feel *more like* themselves, not just *better about* themselves.[68] But if cosmetic surgery is anything to go by, rather than promoting individualism, enhancement could lead to a different kind of uniformity. The universal obsession in the proenhancement literature with a rather superficial measure of intelligence or memory expansion, for example, strikes me as very much like the way in which cosmetic surgery has promoted one cookie-cutter ideal of look-alike beauty.[69]

Like the distinctive if bulbous nose that you inherited from your great-grandfather, your unenhanced self is at least authentically yours: you might not want to change it. What's good about our self-awareness, according to some critics of enhancement, is that it's likewise genuinely ours, no matter how miserable it makes us. By taking Prozac, this argument might run, we're not being authentic to our true selves: instead, we allow ourselves to become alienated from our genuine identities.[70] This existentialist style of critique presumably sees indulging in neurocognitive enhancement as akin to "bad faith."[71]

A less philosophical version of this antienhancement argument is that by using drugs such as Ritalin to control our behavior, we become dependent on a chemical crutch. That too erodes our agency, even when the dependence stops short of addiction. A key component of moral agency is learning to control your impulses.[72] When children who have difficulty learning that process are offered the shortcut of Ritalin, life may become a great deal easier for parents and teachers, but the children themselves are being deprived of the chance to learn how to become better selves—which would be the genuine enhancement, in this argument. On the other hand, use of these drugs, perhaps on a temporary basis, might enable children with attention-deficit disorder to assert more control over their behavior. The argument about moral agency and enhancement can cut two ways.

An extreme example of the way in which selfhood is actually undermined by neurocognitive technologies is the operator (or operatee?) of a brain-machine interface (BMI), for example, a soldier mentally controlling a drone aircraft. "If you are controlling a drone and you shoot the wrong target or bomb a wedding party, who is responsible for that action? Is it you or the BMI?" asks Rod Flower, the chair of a 2012 Royal Society working party on neurocognitive technology and the military. "There's a blurring of the line between individual responsibility and the functioning of the machine. Where do you stop and the machine begin?"[73]

Those favorably disposed toward neurocognitive enhancement view our identities as malleable and capable of improvement through such technologies while still remaining truly ours. As Buchanan puts it in dismissing the argument that enhancement is character deforming: "Instead of arguing that enhancement is too risky to our character, why not proceed in the opposite direction and argue that given how deficient our character is, we may need moral enhancement technologies?"[74] To the concern that "moral enhancement" technologies might actually lead to atrophy of the moral powers, he replies: "Traditional moral education involves 'technologies,' such as rule-following (and, on some accounts, deference to religious authority) that are designed to *replace* moral deliberation about particular matters."[75] While Buchanan falls prey again to his tendency to let enhancement stand for anything and everything, he does have a point. Traditional moral education wasn't necessarily about enhancing authenticity or agency and might even have been all about stamping it out.[76]

Enhancement's advocates see themselves as having embarked on a quest for the Grail of truer selves than the inferior ones we now inhabit. In their view, critics of enhancement are condescending paternalists who want to enforce a standard low-grade uniformity.[77] What right do they have to stand in the way of the noble quest for self-improvement? To this charge one of enhancement's most prominent critics, Michael Sandel, answers:

> I do not think the main problem with enhancement and genetic engineering is that they undermine effort and erode human agency. The deeper danger is that they represent a kind of hyperagency—a Promethean aspiration to remake nature, including human nature, to serve our purposes and satisfy our desires. The problem is not the drift to mechanism but the drive to mastery. And what the drive to mastery misses and may even destroy is an appreciation of the gifted character of human powers and achievement.[78]

If we did succeed in remaking human nature to that extent, could we predict whether it would be for better or for worse? Buchanan acknowledges our poor track record in making decisions that actually benefit ourselves but considers that to be an argument *for* rather than against enhancement. In terms of our manifold cognitive biases and judgmental errors, things can only get better, he thinks.

The difficulty here is that it's our muddle-headed present selves who are in charge of designing the "enhancements." You remember them: the ones whose thinking is so foggy that they need radical help, possibly extending irrevocably even to the genomes of their descendants. They're also the ones who are so prone to look for a technological fix when things go wrong—and enhancement is nothing if not a technological fix. Buchanan appears prey to that reasoning himself, for example when he suggests that one genuine enhancement would be the ability to tolerate more extreme fluctuations of climate and temperature caused by global warming. The obvious retort is that it would be better to make a last-ditch stand against global warming rather than trust in the technological hubris that got us into this mess to get us out again.

There's a similar but perhaps even more troubling question about our inability to predict not just the cognitive makeup of the "transhumans" or "posthumans" who could supposedly be created by such massive interventions but also their moral sensibility. What if rather than being closer to our authentic selves, the enhanced turned out to have a set of values hostile to our own? Critics of enhancement, such as George Annas, stress the likelihood that the enhanced would constitute a powerful new social elite and the risk that they would have very little regard for the unenhanced underclass.[79]

Buchanan notes that we can't know whether this would happen: "Even if biotechnology eventually yields enhancements that are so radical as to call for a new, higher moral status category for the enhanced, the moral status of the unenhanced would not *thereby* be diminished."[80] That's perfectly plausible in terms of our existing concept of human rights as universal, but we can't predict what judgments about moral status the "posthumans" might make. Given that they've been engineered to be "superior," they might not be all that charitable.

GLOBAL JUSTICE, PERSONAL RELATIONSHIPS, AND "MORAL ENHANCEMENT"

Even if "posthumans" turned out to be as nice as pie, some critics argue that creating them would be nothing less than a crime against humanity. George Annas, who even calls it "genetic genocide," says that "inheritable genetic alterations can be seen as crimes against humanity of a unique sort: techniques that can alter the essence of humanity itself by taking human evolution into our own hands and directing it toward the development of a new species, sometimes termed the posthuman."[81] By irrevocably altering the essence of what it means to be human, this argument runs, a unique offense has been committed not just against individuals but against *homo sapiens* as a species. Likewise, in his book *Our Posthuman Future*—whose title seems to accept that "posthumanism" is grounded in real science—Francis Fukuyama argues against enhancement, and seemingly all advances in biotechnology, in these terms: "What is it that we want to protect from any future advances in biotechnology? . . . We do not want to disrupt either the unity or the continuity of human nature, and thereby the human rights that are based on it."[82]

But although the argument from human nature and the natural is common, I agree with Buchanan that it's weak. It's natural for a very large proportion of babies to die before the age of one, but we try to do something about it. For advocates of enhancement to argue that we should likewise try to improve on our natural lifespan or our natural propensity to violence isn't unreasonable in itself. What just comes naturally can be good or bad.

I'm more inclined in favor of two other We-style arguments against the race toward enhancement:

1. Concentrating on enhancement technologies increases *distributive injustice.*
2. Concentrating on enhancement technologies alters *personal and social relationships for the worse.*

Note that unlike Fukuyama's or Annas's arguments, neither kind of claim assumes that enhancement technologies will actually *succeed*: the *attempt* is bad enough. By diverting scarce medical resources from public health measures, for example, spending money on enhancement research increases distributive injustice on both national and global levels. By trying to control our children's genetic makeup rather than accepting them for what they turn out to be, we treat them as commodified objects rather than people. Or so these arguments run. Let's examine them in greater depth.

In the first camp, the World Health Organization report *Genomics and World Health* explicitly links the therapy-enhancement distinction with securing equality of access to health care: "A fundamental part of the moral imperative of health care is its role in maintaining normal function, and in turn helping secure equality of opportunity for persons that serious disease and disability undermine [sic]. Genetic enhancements of normal function, on the other hand, do not serve justice in this way."[83] Likewise, in his book *Just Health*, Norman Daniels argues that placing enhancement on an equal footing with therapy, in terms of resource allocation, worsens injustice for those who have the misfortune to suffer pathologies that impair their normal functioning. To the proenhancement argument that genetic fate affords some people better life chances than others anyway and that enhancement merely builds on that kind of natural difference, Daniels replies that it's not a question of eliminating all individual difference. Rather, the notion of fair shares demands that we should do our best to lessen the ill health and ill chance suffered by those who have to endure conditions that diminish their chances of living a more normal life.[84]

Another reply to the it-happens-already argument might be that enhancement will generally entail devoting scarce resources to those who are already privileged, thus reinforcing or worsening existing inequality. Neurocognitive drugs for wealthier Americans will get priority over the construction of latrines to prevent children's deaths from diarrhea, dysentery, and cholera in Liberia. To say that Liberians are already suffering from distributive injustice—and to imply that they can just suffer some more—is callous and counterintuitive. Liberia might well have a particular claim on our consciences, with its original history as a refuge for freed American slaves and its cooperation during World War II as a U.S. listening station. And given that diarrhea from contaminated water kills more children worldwide than AIDS, tuberculosis, and malaria put together,[85] better sanitation would save a lot of young lives. But the point doesn't just apply to Liberia or to children.

In general, the it-happens-already style of argument is laughably weak: it's a form of what philosophers call the naturalistic fallacy, the illicit jump from "is" to "should." (The argument from human nature is another form of the same fallacy.) Saying a phenomenon exists doesn't tell you anything about whether it should be tolerated. We don't accept that, because murders happen already and will continue to happen whatever laws are enacted, we shouldn't bother trying to outlaw murder.

If we agree with Daniels that we should make it a resource priority to help those at the bottom of the ladder climb up a bit rather than add extra rungs so that those already near the top can look down on the plebs from an even loftier height, we still have a problem: what counts as normal? Daniels acknowledges that the possibility of normal functioning can be affected by conditions such as extreme shortness, which aren't pathologies as such—just one end of the height spectrum. He recognizes that reasonable people could disagree about whether nonmedically indicated use of human growth hormone should be covered by public health or insurance plans but still wants to maintain that it's an enhancement rather than a therapy, even though very short people may be socially disadvantaged. Without the distinction between enhancement and therapy as restoring normal functioning, Daniels thinks, there would be an insatiable appetite for every intervention possible, which not even the wealthiest society could afford for all its members.

Health, understood as the absence of departures from normal health or pathology . . . is a finite or limited concept, unlike income or wealth. This conception matches the way health is viewed in the actual work of medicine and public health, where it is treated as a threshold (or, better yet, a ceiling) that we strive to reach but not exceed. In this regard it is unlike money: We can always have more, without limit.[86]

It looks so obvious as to be trite that spending money on enhancements means *not* spending it on something else, such as public healthcare. And enhancement is certainly open to the charge of creating or reinforcing a limitless appetite for "improvements." But while I'm very sympathetic to the spirit of these arguments, I'm not entirely convinced that enhancement will necessarily take resources away from the poorest.

For a start, the payers aren't the same: national governments and international nongovernmental organizations pay for public health measures like vaccination, whereas enhancement is mainly paid for by individuals and, sometimes, their insurers. Like other forms of Me Medicine that we've examined, enhancement technologies are all about creating demand for a product that you couldn't have imagined and a market where none existed before. That's particularly true of retail genetics and private umbilical cord blood banking, and it would also be true of neurocognitive enhancers, if the evidence base became better. The demand for them would come predominantly from individuals, but corporations would also get very much involved.

Carl Elliott notes that pharmaceutical companies have pushed hard to get doctors to accept products such as Prozac as treatment rather than enhancement, allowing patients to claim reimbursement.[87] Similarly, through product placement in the media, he says that these firms encourage patients to ask their family practitioner for specific brand-name drugs. If neurocognitive enhancers ever broke through the evidence barrier, the same strategy would probably be employed by pharmaceutical firms eager for new markets, as patent protection expires on their mainstream money-printing drugs.

In some cases, the fact that enhancement is mainly an individual consumer choice will indeed lead to inequities: for example, when college students who can afford neurocognitive drugs edge out their less wealthy classmates in the grades stakes and then the jobs market.[88] A permissive policy toward the use of "smart drugs" on campus was advocated in a *Nature* article in 2008.[89] Those authors argued that there were already inequities in access to higher education, so there was no objection to adding a new one—the ability to pay for smart drugs. (You may recognize this as our old disreputable acquaintance, the "it-already-happens" argument.) But while smart drugs on campus do raise issues about distributive justice, that question isn't of the same level of gravity as concerns about public health programs.

The bigger question is whether governments and NGOs will be starved of resources for We Medicine programs because the money will instead be going into enhancement, a Me Medicine concern. While I'm sympathetic to the view that scarce public health resources might be diverted into enhancement, particular at a time of government cutbacks round the globe, I'm not absolutely convinced that the danger is inevitable, even though I think the risk is real. But perhaps it's more psychological and political than economic: if consumers and voters come to see Me Medicine in the form of enhancement as the way of the future, there will be even fewer votes in those latrines for Liberia.

Will the glittering promise of enhancement technologies tempt national leaders to spend their money on them rather than on public health? Although this is an exception to the way in which enhancement would generally be funded from private wallets rather than the public purse, I think the jury is still out on this one. Take the case of military uses for enhancement technologies.

The U.S. military has gone through successive phases of enchantment and disenchantment with enhancement.[90] After a phase of infatuation with "wired" soldiers and networked sensors around 2002, the army changed its tune when it became clear that elite special forces alone couldn't overcome the hostility toward invading forces in Iraq. Approximately $4 billion in annual research funding was shifted from high-tech military enhancement, such as a helmet and vest that monitored the soldier's brain, to mundane but urgent matters such as better protection from roadside bombs.

However, a more recent report suggests that soldiers who had transcranial direct current stimulation as part of a virtual reality training program were able to spot roadside bombs, snipers, and other hidden threats twice as fast as other trainees.[91] These findings, plus the likelihood that weapons manufacturers would see dollar signs flashing in this report, suggest that we might be going into another enchantment phase. But then again, we're also entering an age of public-sector austerity, which means cuts for military spending.

What about the second We-style argument against enhancement technologies—that pursuing them might alter personal and social relationships for the worse? This is hotly contested terrain, with proenhancement writers recently devising a counterstrategy by claiming that our capacity for productive and peaceful relationships could be vastly improved through "moral enhancement."[92] Indeed, some of them assert that without "moral enhancement," cognitive enhancement could actually be dangerous,[93] although that view is roundly rejected by other enhancement advocates.[94]

In Aldous Huxley's *Brave New World*, moral insight has been enhanced as straightforwardly as laser eye surgery can now improve physical eyesight. Rather than wearing their moral beliefs on their sleeves, denizens of the future stow them in their pockets, as bottles of "soma" tablets. As the character Mustapha Mond explains, "Anybody can be virtuous now. You can carry at least half your morality round in a bottle. Christianity without tears—that's what soma is."[95]

But here in our boring old world, hopes for "moral enhancement" rely on neurocognitive transformations that are not substantiated in the medical evidence base. To be more precise, while there is a growing body of work on the way in which neurochemical transmitters such as serotonin and oxytocin favorably affect people's perceptions of fairness and trust in social situations,[96] that research isn't yet at the stage of clinical trials to determine whether these substances could or should be administered as the equivalent of soma tablets. It's not even been established whether such trials are possible in principle, because as one prominent researcher in the area admits, "the precise neural mechanisms underlying those relationships [between neurotransmitters and prosocial behavior] remain unclear."[97]

The neuropeptide oxytocin, released during pregnancy, lactation, and childbirth, is a particular favorite among "moral enhancers,"[98] who point to its supposed role in making people more cooperative. Of the seventy-odd references in the Cochrane Library to meta-analyses evaluating the function of oxytocin, none actually substantiates this purpose. However, there has been military interest in using oxytocin to interrogate detainees and prisoners.[99] This is probably not the use that would-be "moral enhancers" have in mind.

While the lack of an evidence base very much undermines the position of those who think "moral enhancement" is the way of the future, antienhancement arguments about harmful effects on relationships, as noted before, don't depend on whether such enhancements have actually succeeded. It's trying to use them to alter others that's wrong, whether those others are alive now or yet unborn. I think this is a crucial point.

I'm certainly not arguing that enhancement is simply eugenics by another name—even though one proenhancer has defiantly called his book *Liberal*

Eugenics[100]—because I recognize that it's not coercive. The closest that proen-hancement writers come is the assertion that that enhancement is a moral duty.[101] But there is a logical similarity: what's (deeply) wrong about eugenics isn't mitigated by the ultimate failure of large-scale eugenic programs such as the Holocaust or the sterilization campaigns against the "feeble minded."[102] Similarly, the *attempt* at enhancement—particularly transgenerational ge-netic engineering—is wrong, whether or not it's possible or successful, be-cause of its manipulative attitude toward other persons and its baleful effect on our relationships with them. That, at least, is how this second We-style ar-gument might run.

What social solidarity we possess in an individualistic, market-based soci-ety could be put at risk by enhancement technologies, according to Michael Sandel. Before we embark on the enhancement enterprise, he says, we should stop to think that we don't even own the talents we have. Still less, by implica-tion, would we be able to take credit for those we attain through taking drugs.

> If bioengineering made the myth of the "self-made man" come true, it would be difficult to view our talents as gifts for which we are indebted, rather than as achievements for which we are responsible. This would transform three key features of our moral landscape: humility, responsibil-ity, and solidarity. . . . The natural talents that enable the successful to flour-ish are not their own doing but, rather, their good fortune, a result of the genetic lottery. If our genetic endowments are gifts, rather than achieve-ments for which we can claim credit, it is a mistake and a conceit to assume that we are entitled to the full measure of the bounty they reap in a market economy. We therefore have an obligation to share this bounty with those who, through no fault of their own, lack comparable gifts. A lively sense of the contingency of our gifts, a consciousness that none of us is wholly re-sponsible for his or her success, saves a meritocratic society from sliding into the smug assumption that the rich are rich because they are more de-serving than the poor.[103]

Other political theorists, notably John Rawls, regard intelligence as some-thing bestowed on us by fortune, not an attribute for which we can claim merit.[104] Sandel likewise urges us to "recognize that our talents and powers are not wholly our own doing, despite the effort we expend to develop and to exercise them." Acknowledging the "gifted quality of life" teaches us, in the words of the Phil Ochs song, that "there but for fortune go you and I."

So there's a link between this second We-oriented style of argument and the first one, the one about distributive justice. Sandel's position seems appealingly humble—a quality often in short supply on the other side of the enhancement debate. I'm not entirely convinced by it, however, because I agree with Le Doeuff that we don't just receive our intelligence passively: we also create (or undermine) it by our actions. I have misgivings about the genetic determinism in Sandel's declaration that "the natural talents that enable the successful to flourish are not their own doing but rather . . . a result of the genetic lottery." But then Sandel does say that "none of us is *wholly* responsible for his or her success." If he were a full-fledged proponent of genetic determinism, he would have to deny that we are even *partially* responsible for our own success: then it would all be down to our genes. Presumably he must recognize that there are other factors than the genetic lottery behind success, even though he rightly asks us to remember the genetic lottery and its implications for social justice.

Sandel also connects this recognition of the "gifted quality of life" to parenthood, which to his mind is (or should be) the supreme example among human relations of "openness to the unbidden." His argument isn't the same as the one we encountered from Habermas at the start of this chapter: it's not so much that enhancement threatens *autonomy* as that it threatens *relationship*.

> The problem is not that parents usurp the autonomy of a child they design. The problem lies in the hubris of the designing parents, in their drive to master the mystery of birth. Even if this disposition did not make parents tyrants to their children, it would disfigure the relation between parent and child, and deprive the parent of the humility and enlarged human sympathies that an openness to the unbidden can cultivate.[105]

Those who reject Habermas's view[106] can rightly assert that genetic endowment limits *everybody's* autonomy. While genetic endowment sets limits on what we can achieve, it doesn't necessarily interfere with moral agency or freedom itself (except perhaps in the case of someone whose mental capacity is so seriously undermined, for example by genetically linked learning disability or genetically related schizophrenia, as to overturn the normal legal presumption that the decisions of competent adults are legally binding). The parent who sets out to engineer his offspring genetically makes himself "coauthor of the life of another," in Habermas's phrase, and thereby intends to undermine his freedom. But we're all coauthors of each other's lives, not only in the Bedford Falls of the classic film *It's a Wonderful Life* but also in the wider world.

In contrast, Sandel is arguing that what's wrong about transgenerational enhancement is the *manipulative approach to relationships* that it implies. David Wasserman likewise asserts that the issue about enhancing children isn't government regulation versus reproductive autonomy—as it's often portrayed—but whether it's wrong to try to shape the character and capabilities of someone with whom you expect to enter into a particularly intimate relationship, that of parent and child.[107] While Savulescu thinks that parents are duty bound to produce the best children they can, the opposite view is that they're morally wrong even to try to exert such control. That wrongness has nothing to do with whether they fail, although they probably will fail: genetic determinism is inaccurate for all sorts of reasons, the role of epigenetics among them.

In Kantian terms, it's wrong for me to try to use someone else's mind as the rails on which to run my engine—for example, by lying to another person in an attempt to manipulate her into doing what I want. Two of the most influential enhancement advocates, Julian Savulescu and John Harris, are strong utilitarians, so it's perfectly consistent that even in their extensive and detailed writings, they don't confront this Kantian argument about the inherent wrong of trying to manipulate someone else's mind. Yet even from a utilitarian point of view, society's welfare would be undermined if familial relationships and trust were poisoned by the suspicion of manipulation and control.

As a final point in this section, concern about control and manipulation is a separate worry from the frequently voiced fear that allowing parents absolutely free rein to enhance their children would result in one class of "superpersons" and a new *lumpenproletariat* of the unenhanced.[108] Again, that criticism is only telling if transgenerational genetic engineering actually *works*. But even though it doesn't, it's still wrong to enter into parenthood with the kind of controlling attitude that transgenerational genetic engineering would imply. While I don't believe that any of us ordinary parents, not being saints, can manage fully unconditional love, I think we ought to try.

THE FOUR HYPOTHESES

How do the four hypotheses introduced in chapter 1 stack up against enhancement technologies?

First, for military uses of enhancement technologies, perceived *threat* is certainly relevant. But otherwise, enhancement is sold on promise, not threat—threat's seeming opposite, or as I perhaps naively called it in the first chapter, its virtuous twin.

By now you'll have gathered that I think the promises made by enhancement technology aren't so very virtuous, even if they turn out to be empty (as the weak evidence base leads me to suspect, particularly in the cases of transgenerational genetic engineering and "moral enhancement"). Even the attempt is manipulative, posing a threat of its own to how we conceive of the intimate relationship with our children. And distributive justice—another We concern ignored by the relentless Me-ism of enhancement—is also threatened, although perhaps more indirectly than it looks.

Threat is the hidden subtext elsewhere, too, for example, the way in which the (realistic) fear of high debt and no jobs on graduation tempts college students into competitive use of neurocognitive enhancers.[109] This dilemma illustrates a more general point about Me Medicine, which I've already mentioned in chapter 1. If the enhancement technologies ever fulfilled their potential, we all tend to assume that we would be in the class that benefits from their use. (The exception is moral enhancement, where the presumption seems to be that it's the other guy who needs rescuing from his evil tendencies.) But what if we wound up in the unenhanced *lumpenproletariat* instead?

There are other possible threats: would insurers insist that you paid a higher premium if you'd turned down the chance to magic away your failings through the enhancement technologies? Would employers do the same with any health coverage they provide—assuming, of course, that as one of the Great Unwashed and Unenhanced, you could get a job at all? Most of these projected threats are probably as fictional as the Singularity—the moment to which transhumanists look forward, when an enhanced elite will "slip the surly bonds of Earth" and attain the technoutopian equivalent of the Rapture.[110] But they do illustrate the complex interplay between threat and promise in the enhancement technologies.

Second, *narcissism* looks like an obvious explanation for the interest in enhancement, but it's certainly not how enhancement is sold. Like threat, narcissism is a negative, while enhancement's proponents accentuate the positive and sometimes even stress the socially minded over the narcissistic. For example, Savulescu calls his proposal for a form of transgenerational genetic engineering "procreative *beneficence*"—although its effects on women would be anything but beneficent. As with private cord blood banking, if even more speculatively, this conception appeals to parents' altruism, not their narcissism.

Other forms of enhancement, such as "smart drugs," do seem to be the logical extension of the individualistic sense of entitlement that Twenge and Campbell identify in their book *The Narcissism Epidemic*. Yet as we also saw in retail genetics, the clarion call is to personal responsibility—taking charge of your health—although narcissism is also part of the mix. Geneticization means that those who are well can then take credit for it—not just for their superior genes but also for their initiative in counteracting any "inferior" ones. You can take even more credit for enhancing your existing capabilities—or so the proenhancement argument runs.

Third, if social capital is in decline, then your capital in your own body becomes all important, as I argued in the last chapter of *Body Shopping*. If "Me is all I got," then I want it to be the best Me possible. Enhancement fits right in. This is one important way in *neoliberal political policies and corporate interests* help to explain the enhancement trend, and it doesn't just apply to individuals. Science itself is changing its business model from being a public good to becoming a purveyor of private benefits, according to the testimony from John Sulston in chapter 1.[111]

That model is also affected by patent expiration for many of the most profitable blockbuster drugs. The pharmaceutical industry "has been utterly dependent on patents right from the time of its inception."[112] As patents expire on many mass-market drugs, the industry will need to find something else. Enhancement can be seen as the logical culmination of the pharmaceutical industry's turn from the 1980s onward toward biotechnology, culminating in GlaxoSmithKline's $3.6 billion acquisition in August 2012 of the U.S. biotech company Human Genome Sciences.[113]

Undermining the distinction between treatment and "cosmetic neurology" or other forms of enhancement also benefits the pharmaceutical sector.[114] It would allow the industry to create new markets where none existed before, for products no one ever dreamt of. Enhancement makes that seeking continuous and permanent: there's no end to what you need to do to improve yourself, meaning no stop to potential profits. (Well, none except that pesky lack of an evidence base.)

Although they declare such interests in the approved fashion, researchers into enhancement sometimes have an investment of their own in firms such as Memory Pharmaceuticals, which tested a drug to improve memory consolidation[115] and which was subsequently acquired in 2008 for $50 billion by the giant Roche pharmaceutical company. Reports from governmental bodies and investigating journalists document that enhancement projects are backed by funding not only from the pharmaceutical industry but also by information technology companies, including Intel and Google.[116]

According to a *New York Times* report, Singularity University, dedicated to the idea that a combined human-machine intelligence will eventually replace our limited human minds, has numbered Google's Larry Page, Sergei Brin, and many other denizens of Silicon Valley among its supporters, with Google having contributed over $250,000 in donations. The Singularity project can be seen as another form of Google's own target, "building a giant brain that harnesses the thinking power of humans in order to surpass the thinking power of humans."[117] Even where there's no specific investment by any particular company, favorable publicity for enhancement, frequently unencumbered by any scientific evidence,[118] provides soothing background music that puts consumers in a favorable frame of mind toward biotechnology in general.

Finally, perhaps to dispel the (admittedly unfounded) charge of eugenics, enhancement advocates frequently present *choice and autonomy* as their motivating forces. Choosing enhancement supposedly enables you to be true to your authentic self, enhancing your personal autonomy.[119] It allows you, in this formulation, to feel more like yourself.

But the limited neurocognitive enhancements that are already possible, such as off-label prescribing of Ritalin or Adderall, actually *diminish* choice by leading to a race to the bottom. Those college students who can't afford them or who simply can't get access may lose out because they have no other option than to use their own unassisted mental powers.[120] If enhancement ever broke through the evidence barrier and started to become as normal as cosmetic surgery, we'd probably see the neurocognitive equivalent of today's growing phenomenon of job seekers feeling obliged to undergo face lifts and tummy tucks to improve their chances in the employment market.

More generally, enhancement's leading advocates present it as an absolute moral obligation, for example, the duty of procreative beneficence. That doesn't sound a lot like you have a choice. In this respect enhancement is similar in its "branding" to umbilical cord blood banking, which is also portrayed as a moral must.

This chapter concludes my "reality check" of four prominent Me Medicine technologies: retail genetics, pharmacogenomics, private umbilical cord blood banking, and enhancement. In all but pharmacogenomics, and even there to some extent, I've found a striking lack of gold-standard evidence to support the contention that you have no choice but to embrace the personalized medicine "revolution." I've also found corporate interests and neoliberal public policy to be the one explanation that works across all four technologies, although threat comes in a not-too-close second.

Now I want to turn to the opposite phenomenon: the way in which genuine evidence is written off as mere media "hype" and corporate interests actually *over*emphasized in one of the most contentious We Medicine programs of our time: vaccination. Why are the largely unverified promises of Me Medicine taken as gospel truth, while the potentially vast threat of pandemics is dismissed as a fabrication by drug companies to sell their vaccines? You may well ask.

6

"THE ANCIENT, USELESS, DANGEROUS, AND FILTHY RITE OF VACCINATION"

Public Health, Public Enemy?

I find it very interesting that the vaccine does the opposite of what its [*sic*] supposed to do. Is any one open to the thought that this is intentional? That the people in power are using this as a means for population control? And the fact that governments are in the process of making the vaccine MANDATORY?
—Reader comment on article "Swine Flu Jab Link to Killer Nerve Disease,"
Daily Mail Online (UK), August 15, 2009

IT'S CERTAINLY TEMPTING TO DISMISS COMMENTS LIKE THIS AS "crackpot conspiracy views,"[1] but that risks becoming counterproductive. Vaccination has plausibly been called "medicine's greatest lifesaver":[2] it's emphatically not a means for population control. The first and only contagious disease to have been completely eradicated, smallpox, was defeated through vaccination. But in the opinion of Arthur Allen, a leading historian of vaccines, "while vaccination seems to be more efficient and safer than ever before, public ambivalence about the practice has rarely been higher."[3]

Worldwide, popular reactions against vaccines for influenza, childhood diseases, and cervical cancer threaten immunization programs that public health experts see as crucial. Dismissing these reactions as "moonbeams from the larger lunacy," in the words of the Canadian humorist Stephen Leacock, only strengthens the antivaccination movement, which has frequently succeeded in painting its opponents as elitist and condescending. While the comment above may put its suspicion of "the people in power" in particularly strident fashion, the underlying sentiment is common to all the antivaccination movements. Rather than seeing vaccination as a measure of immense communal benefit—the epitome of We-ness—they view immunization as imposed by the state in top-down, undemocratic fashion. Frequently, government is portrayed as acting hand in glove with the pharmaceutical companies that produce the vaccines.

While Me Medicine basks in the glamour of stem cell research, genetics, enhancement, and other twenty-first-century promissory technologies, We Medicine in the form of vaccination is boring old news, it appears. Over a century ago, it was already being denounced as "the ancient, useless, dangerous and filthy rite of vaccination."[4] Perhaps it's also tainted by its origins, in the distinctly unglamorous form of the cowpox-based vaccine developed by Edward Jenner in the late eighteenth century.

Whereas the unknown is a realm of exciting potential to the proponents of enhancement, vaccination and other protective measures against pandemics are premised on the disturbing idea that the unknown is a threat.[5] Pandemics remind us of our mortality in a way that modern biotechnology soothes us into forgetting. They show that the stubborn threats we thought we'd vanquished still impede our transformation into "posthumans."

Historically, resistance to mass vaccination has been the norm rather than the exception. If we need reminding of that fact, it's because for a period in the mid-twentieth century—roughly the same span of time identified by Putnam as the high tide of public-spiritedness—that wasn't the case.

> Antivaccine thinking receded in importance between the 1940s and the early 1980s because of three trends: a boom in vaccine science, discovery, and manufacture; public awareness of widespread outbreaks of infectious diseases (measles, mumps, rubella, pertussis, polio, and others) and the desire to protect children from these highly prevalent ills; and a baby boom, accompanied by increasing levels of education and wealth.[6]

This was also the period when Medicaid and the Early and Periodic Screening, Detection, and Treatment Program were passed, as well as the time when the Centers for Disease Control and Prevention (CDC) robustly championed school vaccination programs.[7] All these We Medicine measures were premised on the idea that government intervention could bring medical benefits— that public health wasn't actually a public enemy.

But by the late 1970s or early 1980s—paralleling Putnam's time scale for the rise of the "bowling alone" mentality—this golden age of public acceptance for vaccination was largely over. One major factor was the notorious "vaccine roulette" fiasco, when concerns about brain damage from the whole-cell pertussis vaccine and threats of litigation led to the withdrawal of licensed vaccines by manufacturers and a wider concern about vaccine safety.[8] At present, we live in the period of "business as usual": antivaccine sentiment is powerful once again. True, it erupts around particular vaccines—smallpox, pertussis, measles (MMR), influenza, polio, and human papilloma virus (HPV)—leaving many vaccines accepted much of the time, but it's now global.

In northern Nigeria, for example, a massive popular boycott of a polio immunization campaign sabotaged efforts to reduce the high incidence of the disease.[9] A substantial factor in the resistance was the legacy of mistrust left from a clinical trial of the oral antibiotic Trovan in 1996, also in northern Nigeria, which ended in a U.S. federal court lawsuit brought by thirty Nigerian families alleging lack of informed consent. Another cause was the statement by Dr. Detti Ahmed, the head of the Supreme Council for Sharia in Nigeria, that polio vaccines were "corrupted and tainted by evildoers from America and their Western allies."[10] The result was a large-scale outbreak of poliovirus serotype 1, which spread from Nigeria to western and central Africa and was later brought via Hajj pilgrims and migrant workers to Yemen, Saudi Arabia, and Indonesia. Over 1,500 children were left paralyzed.

What accounts for such resistance to vaccination in the developed and developing worlds? After all, as the director of the World Health Organization reminds us, "It really is all of humanity that is under threat during a pandemic."[11] But when vaccination is compulsory (directly or indirectly, as a condition of school admission), it curtails individual liberty—as do quarantine, travel restrictions, and other compulsory responses to pandemics. More local factors are also important—the 2003 boycott in Muslim northern Nigeria was aggravated by the U.S. invasion of Iraq that same year, breeding suspicion of any move from the West—but choice and consent are common powerful themes. Indian resistance to HPV vaccination against cervical cancer, which I'll examine at the end of this chapter, turned on the absence of consent from the girls vaccinated and their parents.

Mandatory vaccination is no mere figment of a blogger's imagination. Laws passed in England and Wales between 1853 and 1871 required parents to vaccinate their children against smallpox. But while middle-class families could afford their own doctors, the impecunious had to use doctors appointed under the detested Poor Laws. Official Poor Law Guardians were authorized to seek out and discipline noncompliers, seizing their pitiful assets and forcibly vaccinating their children—against which leading thinkers of the period led a resistance movement.[12] Although it was premised on the "Me" of individual liberty, this cross-class alliance demonstrated the "We" of social solidarity. After English law was amended in 1907 to allow conscientious objection, the resistance movement died out in England and Wales—just as it was starting to gather steam in the United States.

At the turn of the twentieth century, a smallpox epidemic stalked the land—and so did teams of vaccinators, who combed city tenements and inoculated their residents, sometimes against their will.[13] Regulation was left to individual states: only about one-third had compulsory vaccination laws, but those that did, such as Massachusetts and New York, were often those with the highest immigrant populations (including African Americans who had emigrated northward after the Civil War). While in England the victims of forcible vaccination were largely defined by class, unsurprisingly in the United States ethnicity also entered the equation. A bacteriologist boasted that an outbreak in St. Louis had been stamped out when he and his colleagues "vaccinated the whole male negro population of the city, and as many women as could be captured."[14] Border officials were empowered to vaccinate forcibly any Mexican immigrants who didn't have a recent vaccination scar—even though Mexico actually had better rates of smallpox vaccination at that point than did the United States.

In November 1901, it emerged that over ninety people had died of lockjaw a few weeks after they were vaccinated, because of unsterilized instruments and batches of vaccine contaminated with tetanus bacilli. Meanwhile, a new and much less lethal variant of smallpox, *Variola minor*, had established itself as the dominant form of the disease in the United States. In 1906, slightly fewer people died of smallpox—ninety out of 15,223 recorded cases—than had succumbed to the contaminated vaccine five years earlier. "Smallpox began to seem less of a menace than the means of preventing it"[15]—a balancing judgment that people also made, rightly or wrongly, during the "swine flu" pandemic of 2009.

By 1908, the Anti-Vaccination League of America had been established, dedicated to the principle that "no State has the right to demand of anyone the impairment of his or her health."[16] Although the patriotism of two world wars and successful campaigns against other diseases like diphtheria, polio, and tuberculosis later gave public health medicine a legitimacy it hadn't previously enjoyed—resulting in that high tide of provaccination sentiment after World War II—vaccine skepticism never died out entirely.

The themes of the antivaccination movement haven't changed greatly over the years: government intrusion—consciously or unconsciously harkening back to mandatory policies—pitted against individual and parental autonomy.[17] Distrust of the medical profession also seems to persist, from the bad old days of the English Poor Law doctors and the armed mob that drove public health officers out of Georgetown, Delaware, in 1926. What's new is that antivaccination movements are now well-orchestrated media and social networking campaigns by successful single-issue groups—using the tools of We-ness to spread a message that threatens one of the cornerstones of We Medicine. In his hard-hitting book *The Panic Virus*, Seth Mnookin spells out how he thinks people invent their own versions of the scientific evidence:

> Combined with the self-reinforcing nature of online communities and a content-starved, cash-poor journalistic culture that gravitates towards neat narratives at the expense of messy truths, this disdain for actualities has led to a world with increasingly porous boundaries between facts and beliefs, a world in which *individualized* notions of reality, no matter how bizarre or irrational, are repeatedly validated.[18]

It might seem perfectly reasonable to say that the decision whether or not to vaccinate a child should be a matter of individual parental autonomy. However, the diseases targeted by vaccination are generally contagious, which casts a different light on whether it's only the individual who counts. Tetanus is not a contagious disease, so inoculation protects the child alone.[19] But that's not true of vaccination for measles, polio, whooping cough, chickenpox, mumps, meningitis, hepatitis, diphtheria, pneumonia, and influenza. If I choose not to be vaccinated or not to have my child inoculated, my child and I won't be the only ones who suffer if we catch the disease and then pass it on to someone else. Autonomy can't be the only consideration with these diseases: Me and We are inextricably interwoven.

By choosing not to have myself or my child vaccinated, I also decrease "herd immunity" and increase the pool of people in whom the disease can take up residence before spreading to others—jeopardizing infants below the age of safe vaccination and elderly people, for whom vaccination is less likely to work. Since the poorest ethnic groups have the lowest levels of vaccination,[20] they are also put at unfair risk when wealthier groups refuse immunization and so decrease herd immunity. And it is primarily the better-off we're talking about: they have a higher success rate in obtaining exemption certificates from the vaccines children need to have before school admission.[21]

It's all very well for antivaccinationists to say that "Our nation can tolerate a certain percentage of unvaccinated children without risking the overall public health in any significant way. Since most children are vaccinated, our nation has enough 'herd immunity.'"[22] If everyone followed this "free rider" logic, of course, no children would be vaccinated, and herd immunity would disappear.

In facing a possible pandemic of contagious disease, we're all both vectors and victims: potential carriers of infectious disease and possible targets for it.[23] Vaccination against the H1N1 influenza virus, which will be discussed at greater length in the next section, illustrates both aspects. Not only does it reduce the patient's own risk of infection, but it also reduces the odds of transmission to others.[24] It's both prudent and altruistic.

What I lose in liberty by agreeing to be vaccinated or to restrict my movements during a pandemic, I hope to gain by simply staying alive. Although popular opinion and some experts[25] view vaccination as an altruistic sacrifice, it's actually prudent self-saving behavior: obviously I stand to benefit if the disease is stopped in its tracks before I become infected. So it's not actually a question of We trumping Me: I won't be Me at all if I die in the pandemic. And for Me, read We, because that self-evidently applies to all of us.

> Constraints do not simply wrong someone by restricting him, yet do right by protecting others. Constraining (and thus from one perspective apparently wronging) given individuals is also doing right to these same individuals, by reducing their vulnerability to the mutual transmission of disease in the interconnected web of human biological/ecological relationships. . . . The issue is not "me against you," but "everyone in this together."[26]

Even the slightest probability that I might die if I remain unvaccinated against pandemic disease would dictate that I should follow the strategy known to philosophers as Pascal's Wager (see chapter 4). On a Pascal's Wager argument, it's rational to be vaccinated even if the probability of a pandemic appears low, because no matter how low the probability, it's cancelled out by the threatened loss of your life, which is effectively an infinite loss: without your life, clearly no other gains can be enjoyed. However, that reasoning might not hold if the possibility of fatal side effects from the vaccine is higher than the pandemic risk.

For example, the *Daily Mail* article that triggered the blogger's comments at the start of this chapter[27] refers to the side effect of Guillain-Barré syndrome (GBS), a subacute, progressive muscle weakness.[28] During a previous "swine flu" outbreak in 1976, there were five hundred incidences of GBS and twenty-five fatalities, out of forty-five million people vaccinated.[29] The Institute of Medicine concluded that there was a causal relationship in adults between receiving the vaccine and developing the condition. However, subsequent studies didn't confirm that conclusion, and vaccination manufacture techniques had also changed by time of the second swine flu outbreak in early 2009. By November of that year, fewer than ten GBS cases had developed out of sixty-five million people vaccinated.[30]

A New York Academy of Medicine survey in 2004 revealed that many more people were indeed worried about the side effects of smallpox vaccine than were concerned about the disease itself: twice as many, in fact.[31] That may well be because we live an age where it is apparently possible to believe honestly that "the scourge of ageing is worse than smallpox."[32] Few of us have ever seen someone die of smallpox, as did the nineteenth-century American writer Margaret Fuller: Seth Hasty, the captain of the ship on which she was returning from Italy to New York, succumbed to the disease before they had left Gibraltar. Although Fuller had been present at many deaths, having served as a nurse in a Rome hospital during the Italian revolution of 1849, she wrote: "I have seen . . . great suffering, but nothing physical to be compared to this, where the once fair and expressive mould of man is thus lost in corruption before life is fled."[33] Fuller herself would die the following month, an indirect victim of smallpox. With the captain dead of the disease, the first mate, an inexperienced navigator, wrecked her ship off the shore of Fire Island. She, her husband, and their one-year-old son Nino all drowned.

In our time, the comparative docility of infectious diseases such as smallpox has created a degree of complacency about the magnitude of the risks. This is the "vexing paradox" about vaccines: "The more effective they are, the less necessary they seem."[34] There's a vexing contrast here with one Me Medicine measure, the private banking of umbilical cord blood: although professional bodies have been pointing out its ineffectiveness for over a decade,[35] it still seems more and more necessary to many parents and grandparents.

On the other hand, public skepticism about smallpox vaccination at the time of the 2004 survey might have been partly attributable to the political use made by the Bush administration of the supposed smallpox threat from Iraq as a justification for invasion.[36] There's a dark side to We Medicine in government manipulation of public fears of "bioterrorism."[37] Another unpalatable truth for advocates of We Medicine lies in the history of quarantine's use against minority ethnic communities.

During the plague scare of 1900, the Chinese community in San Francisco was singled out for quarantine, although the measure was eventually struck down by the Supreme Court. Likewise, during the 1878 yellow fever epidemic in the southern states, armed guards hired by the town forcibly halted trains into Jackson, Tennessee, and detained passengers in improvised fever camps.[38] Whereas in seventeenth-century England the Derbyshire village of Eyam chose to isolate itself to protect others, at the cost of more deaths among the villagers, in San Francisco and Tennessee it was strangers who were targeted against their will. So-called crackpot reactions against mandatory antivaccination measures need to be viewed in their historical and political context of distrust and suspicion.

That history has been sketched out—inevitably all too briefly—in this introductory part of the chapter. Now I want to tell three separate vaccine stories about "swine flu," childhood vaccines, and the HPV vaccine against cervical cancer, with the last case study showing the global range of vaccine resistance. It's in the developing countries—where you might least expect it—that some of the strongest resistance movements to vaccination have arisen. Throughout the world, solidarity has been demonstrated in the struggle *against* vaccination, the very measure that many public health experts would regard as the epitome of We Medicine. What explains this paradox?

Let's begin with the H1N1 influenza outbreak of 2009. Despite the hostility evinced in the quotation with which this chapter began, apathy rather than antipathy was the general response to the "swine flu" epidemic. It's not until we reach the second case study in this chapter, MMR vaccination, that we see public health actively portrayed as a public enemy.

THE 2009 SWINE FLU INFLUENZA PANDEMIC

Poland and Jacobson write:

> The H1N1 influenza pandemic of 2009 and 2010 revealed a strong public fear of vaccination, stoked by antivaccinationists. In the United States, 70 million doses of vaccine were wasted, although there was no evidence of harm from vaccination.[39]

In April 2009, a novel form of the influenza virus was identified in Mexico, causing fifty-nine deaths in Mexico City alone.[40] Although the virus was initially traced to residents of a village who had contracted it from their pigs—hence the nickname "swine flu"—its transmission wasn't limited to peasant farmers. Within a few weeks, the infection had spread to the United States; within two months, on June 11, it had been classified by the WHO as the top ranking of epidemic disease, level 6, a pandemic. By midsummer 2009, the President's Council of Scientific Advisers was predicting that an expected second wave of the virus in the fall could infect 30 to 50 percent of the population, lead to 1.8 million hospital admissions, and cause between thirty thousand and ninety thousand deaths.[41] In the United Kingdom, a "reasonable worst case" scenario of 65,000 deaths was promulgated by the government.

An earlier study, led by Christopher Murray of the Harvard School of Public Health and published in the *Lancet*, had predicted that pandemic influenza of the 1918 "Spanish flu" variety might kill up to sixty-two million people worldwide, given modern population levels.[42] True, the virulence of the H1N1 epidemic wasn't expected to equal that of Spanish flu, which resulted in an estimated fifty to one hundred million deaths around the world. That's more people in one year than the Black Death killed in a century, more victims in twenty-four weeks than AIDS has mustered in twenty-four years, and more fatalities than the number of men who fell on the battlefields of the First World War.

However, Murray's estimates were criticized as excessive and counterintuitive, given that income levels in the developed world now far exceed those in 1918, so that the public's health is generally better.[43] Peter Doshi, who has compiled a database of monthly influenza-related deaths from 1900 to 2004, has found an overall substantial decline in mortality, even taking the 1957–1958 and 1968–1969 pandemics into account.[44] In his study, the 1918 pandemic emerged as very atypical.

The global death toll of the 2009–2010 influenza pandemic was in fact between 72,000 to 162,000, with 2,117 deaths formally confirmed by laboratories in the United States but actual deaths reckoned at between 8,870 and 18,300 by the Centers for Disease Control.[45] Of course, the planners couldn't benefit from 20/20 hindsight—and that's still a lot of deaths. Additionally, the virus claimed a disproportionate number of young adults, so that the total life-years lost were far more substantial than the raw numbers suggest.

There was still no specific swine flu vaccine available in June 2009 when the pandemic was officially declared, only ordinary seasonal flu vaccine—which didn't fit the genomic profile of swine flu—and neuraminidase inhibitors, a class of antiviral drugs targeted at the influenza virus, such as Tamiflu (generic name oseltamivir) or Relenza (zanamivir). Substantial disagreement persists about the benefit of these drugs: although this particular strain of influenza seemed susceptible to antivirals,[46] a Cochrane review found the drugs largely ineffective in preventing or treating influenza symptoms in adults.[47] Takeup of Tamiflu as a prophylactic, particularly for schoolchildren, was hampered by reported side effects including nausea, stomach pain, and sleep problems.[48]

I suggested before that vaccination has the misfortune to look like a tired old technology, and the flu vaccine is no exception. Traditionally, it was made using a fifty-year-old process in which the live virus is injected into fertilized chicken eggs at eleven days old, with the embryos removed. Once identified, the virus has to be incubated in the eggs, extracted, attenuated, and then tested for safety; no more than two doses are usually obtained per egg, and whole batches can be lost through contamination.[49] Because of the long lead-in times in the production of influenza vaccines and because the vaccine has to be specific to the particular strain of influenza, it wasn't possible in 2009 to stockpile requisite quantities of vaccine ready for use as soon as the outbreak started. Only with the December 2011 opening by Novartis in North Carolina of a new vaccine mass-production facility using animal cells in culture rather than chicken eggs did the picture begin to change. Cell culture facilities have now also been introduced by other manufacturers.

This was the first pandemic that had an observed start location and date—although it was identified retrospectively—and drug companies were able to start vaccine production comparatively quickly. All the same, they were tempted to make promises about production dates that they were unable to keep.[50] So even by autumn 2009, with a second peak expected as winter approached, much less vaccine had been produced than was thought likely to be required. That was worrying, because the second peak was forecast to affect 63 percent of the U.S. population.[51]

That failure seems to undermine charges that the flu scare had been drummed up by the drug companies to create more demand for their products: they weren't even able to satisfy the demand that already existed. Several vaccine manufacturers had already ceased production of flu vaccine altogether, because the laborious method made it unprofitable.[52] Diverting limited production resources to a new strain required the remaining companies to gamble most of their resources on predictions about the virulence of swine flu, which meant that fewer doses of standard winter flu vaccine could be produced, an unattractive sacrifice for manufacturers.[53]

All the same, there were allegations of undue influence in the WHO's decision to class the outbreak as a pandemic. Writing in the *British Medical Journal* the year after the pandemic began, Deborah Cohen and Philip Carter uncovered evidence that key scientists advising the WHO on planning for the pandemic had an undeclared conflict of interest: they'd done paid work for the drug firms that stood to profit.[54] Even if the drug companies turned out to have no spare capacity for vaccine production, Cohen and Carter allege that they profited from billions of dollars' worth of government contracts for stockpiled antivirals. This exposé continues to be influential: it was extensively cited in a prominent 2011 article in the *New York Review of Books* by Helen Epstein: "Flu Warning—Beware the Drug Companies!"

Once vaccines finally began to be available in fall 2009, the swine flu epidemic was already quite far advanced, leading to a general perception that there was no point in prevention. "There was a Jekyll and Hyde quality to the risk communication that took place, where it was difficult to obtain the appropriate balance between reassuring people and asserting the need to vaccinate."[55] In particular, there was a conflict between playing down popular fears and explaining that vaccination was still important because flu was killing many young adults. Perhaps the experts didn't succeed in striking the right balance, sending out a mixed message regarded as mealy mouthed and suspicious by the public.[56] They could have taken a stronger line: it was already known that almost two-thirds of those admitted to hospitals in the United States with suspected swine flu were ending up in intensive care.[57] Or perhaps it wouldn't have done any good to take a stronger line, because the media simply doesn't understand the probabilistic nature of forecasting the course of infectious disease.[58]

Either way, by the time the vaccine was ready, the media had already decided that the story—and the pandemic—were over. Since the virus was largely transmitted through schoolchildren, summer vacation had produced a big drop in the number of cases—and in any case, flu is more of a winter disease. By late summer 2009, U.S. and UK media had firmly fixed on the epidemic as being just "hype"—which of course raises the question of who'd been doing the original "hyping."

Although the large numbers infected in the spring had been extensively covered, an equally high second peak in October to December 2009 went almost completely unreported. (You probably didn't hear about the second peak of the outbreak: I didn't either—not until I happened to go to a specialist conference of influenza epidemiologists in London the following year.) People were more likely to consult their family practitioner during the spring peak, partly because they'd read the media coverage and worried that they had the reported symptoms. So it's hard to know whether the second peak might even have been higher without the vaccine,[59] and—crucially—whether although vaccine use wasn't as high as expected, the numbers who did get vaccinated were sufficient to prevent many more people contracting the disease that winter. That would be the expected effect of herd immunity.

In the end, as we've seen, seventy million doses of vaccine went unused in the United States alone—about the same number wasted there as used worldwide, with something over sixty-five million doses administered in sixteen countries. That doesn't prove either that the epidemic was a flash in the pan or that the value of the vaccine was exaggerated, although over two-thirds of the respondents to one survey agreed that the pandemic was all hype.[60] To many people in the United Kingdom and the United States, the swine flu pandemic became just one more instance of overintrusive and bumbling government authorities crying wolf—adding yet another chapter to the long history of public hostility to vaccination.[61]

In fact, we simply got lucky: the swine flu pandemic could have been far more serious.[62] This particular variant of the influenza virus wasn't highly pathogenic, and it didn't mutate into a more dangerous form. Influenza viruses possess an unusually strong power to do so, compared to more stable viruses such as measles or chickenpox, and to circumvent the immunity that populations develop to other variants of flu.

While most viruses have a single particle of genome, the genome of influenza viruses is divided into eight pieces, which can recombine into offspring viruses.[63] This mechanism provides the source of the variations that determine the host range, pathogenicity, and transmissibility of influenza viruses. If two different strains of influenza virus infect one cell, it may result in new offspring viruses to which human populations have no immunity. This danger is particularly marked when a genetic component from another species, such as birds or pigs, enters the mix. Then there is a risk that the mutated form of influenza, containing some of the genes that cause illness in other species but with a new power to infect humans, will be transmitted back to human populations. The 1918 influenza was probably the product of this sort of "antigenic shift."

So far as the recent pandemic H1N1 strain goes, the swine flu virus hasn't mutated this way: with a mortality rate of only 0.026 percent,[64] it lacked the lethal qualities evident in the H5N1 strain of avian influenza A that broke out in poultry markets in Hong Kong in 1997. Although it was initially thought to be confined to birds, by 2008 this strain of avian flu had accounted for 359 documented human infections and 226 deaths, a fatality rate of two-thirds in reported cases.[65] Even the 1918 influenza pandemic killed "only" 2 percent of those it infected. But this strain of avian flu is not transmitted effectively between humans—only indirectly, through intermediate contact with domestic poultry or wildfowl.[66] Nevertheless, a serious threat remains:

> The high pathogenicity of the H5N1 virus for birds and for the few known infected humans was particularly disturbing for local, national, and international public health officials who saw the potential of this virus to create a pandemic similar to the 1918 pandemic. . . . If H5N1 also spreads rapidly from human to human, as did the 1918 H1N1 virus, the result could be devastating.[67]

In late 2011, two sets of researchers, one at the University of Wisconsin and the other at the Erasmus Medical Center in Rotterdam, discovered a mechanism by which avian influenza of the H5N1 strain might spread *directly* between humans in the same manner as the less lethal swine flu. Working with ferrets, in which human flu transmission can be modeled, the Rotterdam researchers found that it required only five genetic mutations to alter avian flu into a disease that could spread directly from animal to animal. The U.S. National Science Advisory Board for Biosecurity took the rare action of requesting that the researchers strike key details from their papers, which were already in press at *Science* and *Nature*. A *New York Times* editorial on January 7, 2012, headlined "An Engineered Doomsday," actually called for the scientists to destroy the transformed virus they'd created.

> We nearly always champion unfettered scientific research and open publication of the results. In this case it looks like the research should never have been undertaken because the potential harm is so catastrophic and the potential benefits from studying the virus so speculative.

While proponents of the research claim that creating the directly transmissible form of the virus would allow a vaccine to be developed and stockpiled in advance, others counter that in nature the virus is unlikely to mutate in precisely the same way as it did in the labs, meaning that any vaccine developed in advance still might not work. And, they contend, the risk of the results or the virus itself escaping into the wrong hands is just too great.

That same month, the researchers voluntarily declared a sixty-day moratorium on publication of the results or further research, illustrating the extreme gravity of the threat.[68] The moratorium was continued by scientists meeting the following month at the World Health Organization, but with the recommendation that eventually full details of the research should and could be published.[69] In June 2012, the research results were published in the journal *Science*. Several prominent scientists disassociated themselves from the decision to publish, which also contradicts the decision of the U.S. National Science Board for Biosecurity.

That lack of consensus among researchers on such a vital question mirrors the discouraging lack of community and solidarity illustrated more generally in the swine flu epidemic. Pandemic influenza is one of the strongest challenges to the Me Medicine view that it's up to me to protect my own health. Yet the overwhelming response to outbreaks of pandemic influenza, even if not as hostile as the comment with which this chapter opened, has often been divisiveness within countries and between nations.[70]

At the international level, poorer countries have refused to share virus specimens unless they're guaranteed equal shares in any benefits.[71] During the avian influenza A (H5N1) outbreaks in late 2006, Indonesia declined to produce virus specimens for the WHO, because of its fear that pharmaceutical companies would use these specimens to patent vaccines and antiviral medications that Indonesia couldn't afford. Given the restrictive history of patent policies,[72] that's certainly not an unreasonable concern, but noncooperation of crucial countries hampers reporting of outbreaks and impedes the development of antidotes.

Under WHO arrangements, the pharmaceutical industry is meant to pay half the annual operating costs of the Global Influenza Surveillance and Response System. In return, the industry gains access to biological materials from Third World countries. But the framework doesn't require richer member states to provide specific levels of vaccine donations to poorer countries. Vaccines weren't equally shared during the swine flu outbreak of 2009, further deepening distrust between the First and Third worlds. Yet "global cooperation and fair allocation of life-saving resources are essential for an effective and humane response to global health threats."[73]

In the United States, people of color have long been underserved by vaccination programs. After the measles vaccine was licensed in 1963, the disease declined quickly among middle-class white children but became disproportionately concentrated in inner-city ghettoes.[74] One study found that only 48 percent of adult blacks and 49 percent of Spanish-speaking Hispanics had received seasonal flu shots, compared to 67 percent of white Americans.[75] When media and public opinion prematurely concludes that vaccination is useless and epidemic diseases nothing to worry about, as happened during the 2009 swine flu epidemic, it undermines efforts to spread the resource of vaccine more equitably among ethnic groups.

At the clinical coalface, it's also worrying to find low rates of vaccination among those who care for the frail elderly. Up to 90 percent of influenza-related mortality occurs in the oldest age groups, but care workers in homes for the elderly have rates of vaccination as low as 5 to 10 percent. (Speculatively, there might be a correlation between that statistic and low vaccination rates in Hispanic and African American communities, if care workers are disproportionately recruited from lower-income ethnic groups.) Two Cochrane reviews demonstrated that if those rates could be increased to between 40 and 50 percent—it doesn't have to be universal—deaths among elderly patients would fall by a statistically significant number.[76] Even if all the residents of care homes were themselves vaccinated, the weaker immune systems of the elderly means that only about 50 to 70 percent of those vaccinated would be protected, compared to the usual rate for adults of 70 to 90 percent.[77] So vaccination for care home workers is the most effective way to protect those for whom they care.

The rest of us can try to avoid public transport or crowded places during a flu outbreak; care home residents can't usually choose to move to reduce their risk. Nor can they choose whether to be looked after only by vaccinated caregivers. So should care workers be obliged to be vaccinated in order to protect their patients?[78] Requiring individual workers to get vaccinated might seem unpalatable, given the long history of hostility to compulsory vaccination, but perhaps employers should be compelled to provide vaccination for those employees who want it, even if they're not covered by workplace health insurance. If it did become mandatory for healthcare workers to be vaccinated, they should also be compensated for any side effects, on the principle that they are victims as well as vectors.[79]

Such important policy questions just weren't debated during the 2009 swine flu pandemic, because the public and the media concluded prematurely—and against the weight of scientific evidence about the possible harms of a pandemic—that vaccination was a tempest in a teapot. In the next example, the opposite occurred: risk was exaggerated, against the medical consensus.

CHILDHOOD VACCINATIONS AND AUTISM

Mark Blaxill, the vice president of the antivaccine group Safe Minds, says, "The damage caused by vaccine exposure is massive. It's bigger than asbestos, bigger than tobacco, bigger than anything you've ever seen."[80]

A few years ago in California, $10 million had to be spent to contain an outbreak of measles, which began with an unvaccinated child and then spread through a pediatrician's office, two supermarkets, and two schools, affecting forty-eight children who were too young to be vaccinated. Although it's one of the most infectious and deadly of all contagious diseases, measles had been nearly eliminated in the 1990s, before a backlash in both the United States and United Kingdom against the measles, mumps, and rubella (MMR) vaccine because of fears that it was linked to autism. Levels of vaccination are now returning to the approximately 90 percent rate needed for herd immunity.[81] But worrying outbreaks continue to occur, for example among students who weren't vaccinated as small children when mistrust of the MMR vaccine was at its height. The World Health Organization has had to defer its expected date for eliminating the disease to 2015, and even that now seems unlikely.[82] This is the narrative I'll tell next, but before I begin, I want to offer a word of caution.

At the beginning of this chapter, I made a tactical argument against dismissing all forms of antivaccinationism as "crackpot conspiracy theories": it's too condescending, if not downright insulting. In this section, I want to go further, to recognize that the parents who were caught up in the brouhaha about a supposed link between childhood vaccinations and autism weren't irresponsible rumormongers but concerned mothers and fathers. Many people who got involved in the movement were parents of autistic children, hungrily seeking an explanation for how and why their previously healthy infants had developed pervasive developmental disorders, apparently after receiving vaccinations. On the other hand, it's equally important to recognize the resentment of what may well be the majority of parents with autistic children—who *don't* accept that vaccines caused their child's condition—against the way in which the media portrays *all* such parents as hostile to vaccination.[83]

I also recognize how daunting the recommended vaccine schedule is, and I fully acknowledge the reasons for feeling that there has been "immunization creep" over the years. The most recent recommended immunization schedule from the Centers for Disease Control, the American Academy of Family Physicians, and the American Academy of Pediatrics begins with injections at the age of two months for rotavirus, diphtheria, haemophilus influenzae type B

(HiB), pneumococcus (PCV), and polio, all to be repeated at the ages of four months and six months. Between six and eighteen months, the schedule adds MMR, varicella, and hepatitis A, with further injections before starting school. Another schedule kicks in at seven years and carries on through the age of eighteen, adding immunizations for tetanus, diphtheria, pertussis, and human papilloma virus (HPV) and the meningococcal conjugate vaccine. Not all the vaccines are recommended for all children, but it's pretty overwhelming all the same.

Although both adult and child vaccination raise the moral issue of population immunity and so transcend personal autonomy, ethically there is a substantial difference between rejecting treatment on your own behalf and rejecting it on behalf of someone else. But that cuts two ways: you could argue that parents have a duty of care that requires them to abide by the balance of scientific evidence and to proceed with vaccination. Or you could recognize that it feels like an assault to expose your healthy child to an inoculation that may possibly do harm, even though it's overwhelmingly likely to do good.

What would Pascal's Wager suggest? Does the consequence of my child developing autism through my own intervention outweigh all the probabilities against it? Does it make a difference that my child was well at the time of the vaccination—although at higher risk of developing a childhood disease if not vaccinated? Although all medical procedures have side effects, they're more typically performed on the sick than on the healthy.[84] Vaccination is different. Even though they view vaccine resistance as irrational, the authors of "a taxonomy of reasoning flaws in the anti-vaccine movement"[85] acknowledge that the risks of vaccination feel real, whereas the harms from not vaccinating feel more hypothetical.

Is it worse to expose your own child deliberately to the vaccine intervention than to do nothing and hope she'll be spared the disease? Or is that a form of parental neglect and social irresponsibility? Is the fate of my own child rightfully of greater concern to me than the question whether measles cases will increase overall?—although, of course, that might also affect the fate of my child. Perhaps the tragic nature of these choices helps explain the extreme emotions evident in the controversy around childhood inoculations, whereas in the swine flu example, it was more a matter of public apathy. Of course, in the flu case, there were also some extremely hostile comments, such as the one with which this chapter opened, but they were isolated occurrences compared to the scale of the anti-MMR movement.

The story of the reaction against childhood vaccines begins with a paper published in 1998 in the *Lancet* by Andrew Wakefield and his colleagues.[86] (Actually, the narrative starts in August 1997, with a prepublication release of the article, which attracted much attention.) The *Lancet* paper argued that the MMR vaccine, in use in the United States since the early 1970s and in Britain since the 1980s, could be responsible for a surge in autism diagnoses. The intermediate link was ostensibly between intestinal disease and autism.

Wakefield and his colleagues wrote that the measles virus had been discovered in the intestinal tracts of patients with irritable bowel syndrome. They hypothesized that some children's immune systems couldn't handle the three-part MMR vaccine all at once, allowing the measles component to become lodged in the lining of the small intestine and making the walls of the bowel porous. The next step, according to Wakefield et al., was that opioid peptides naturally produced during digestion escaped through the newly porous walls, breached the blood-brain barrier, and produced autism through opioid excess in developing children's brains.[87]

In most of the (mere) twelve cases cited, however, autism had *preceded* bowel problems, so there couldn't have been a connection. As critics pointed out at the time, any effects Wakefield and colleagues had observed were at most coincidental rather than causal.[88] The paper, opponents said, was relying on the memories after the event of parents who had come to Wakefield not because he was a pediatrician or a vaccine specialist—he was a gastroenterologist—but because he was known to have an interest in connecting the MMR vaccine with autism, creating insuperable problems of selection bias. Wakefield countered that his critics were simply sticking to conventional medical opinion, ignoring the needs and practical knowledge of parents:

> Hitherto, this claim [of a link between MMR vaccine and autism] had been rejected by health professionals with little or no attempt to investigate the problem. The parents were right. . . . This is a lesson in humility that, as doctors, we ignore at our peril. In many cases, the parents associated onset of behavioural symptoms in their child with MMR vaccine. Were we to ignore this because it challenged the public-health dogma on MMR vaccine safety?[89]

In his continuing campaign to connect the MMR vaccine with autism, Wakefield succeeded to a considerable extent in portraying himself as the underdog, the valiant lone maverick against the medical establishment and the drug companies that made the vaccines—a story adopted enthusiastically by the newspapers. What only came out much later—when huge numbers of parents had already turned their backs on the MMR vaccine—was alleged evidence of his own commercial involvement, uncovered by a freelance investigative reporter for the London *Times*, Brian Deer. Nine months before the *Lancet* article appeared, Deer charged, Wakefield had filed a patent laying claim to "a new vaccine for elimination of MMR and measles virus and to a pharmaceutical or therapeutic composition for the treatment of IBD (irritable bowel disorder) . . . and regressive behavioural disease."[90] Wakefield has since made counterallegations that Deer is not to be trusted because he was writing in a newspaper belonging to the Murdoch organization News International.[91]

When these allegations about Wakefield's financial conflicts of interest were revealed, ten of his twelve coauthors withdrew their names from the *Lancet* paper,[92] and the journal issued a partial retraction. But by then the paper had sparked a U.S. congressional subcommittee inquiry into vaccine safety, beginning in May 1999 and continuing, in full media glare, through the end of 2002. Moved by the situation of his two-year-old grandson, who had been diagnosed as autistic, Chairman Dan Burton "subjected government scientists to harsh cross-examination," charging that "the public-health people have been circling the wagons to cover up the facts."[93]

When another committee member, Henry Waxman, called for a more evidence-based approach, he was booed in the hearing room.[94] Paul Offit, a pediatrician and vaccinologist who was one of the most prominent advocates of the MMR vaccine, received "a lot of hate mail" and even death threats.[95] During the hearings, Offit was asked by Massachusetts Congressman John Tierney whether he had vaccinated his own children. He replied that he had, giving their names and ages. At the next break, an aide rushed up to him, grabbed his arm, and warned him, "Never mention the names of your own children in front of a group like this."[96]

Even reporters striving for balance inadvertently helped fan the flames by adopting a "he said, she said" style of reportage. When a Dallas law firm that specialized in class-action lawsuits teamed up with parents' representatives, more claims began to be filed for autism than were normally lodged for every other type of injury combined.[97] In this poisonously charged atmosphere of what rapidly became the MMR vaccination wars, the fact that subsequent epidemiological studies had disproved Wakefield's contention went unnoticed.[98]

In the United States, however, the mechanism by which vaccines were meant to have caused autism wasn't the link to bowel problems that Wakefield and his colleagues had hypothesized: rather, it was a supposed connection to a derivative of mercury, thimerosal, used as a preservative in vaccines. Although thimerosal was a form of ethylmercury, which is much less toxic than the methylmercury that had caused deaths from Minamata disease in Japan, vaccine opponents argued that the rising number of vaccinations recommended for children meant that they were receiving a highly toxic dose in total. Perhaps public health officials didn't respond to these charges as effectively as they should have done:

> Regardless of the relative harmlessness of ethylmercury, there was a troubling self-assuredness that underlay [U.S.] health officials' blasé attitude to thimerosal's potential risks. . . . Pharmaceutical manufacturers and immunologists at times acted as if the fact that vaccines had saved more lives than any other single invention rid them of the responsibility of ensuring that the vaccines already on the market were as safe as possible.[99]

Antivaccinationists, convinced that drug company interests lay behind this see-no-evil attitude on the part of officialdom, saw their suspicions apparently confirmed when a provision was surreptitiously inserted into the Homeland Security Bill to shield Eli Lilly, which manufactured thimerosal, from vaccine-related lawsuits filed by families.[100] Recent studies, however, generally conclude that there is no evidence that vaccines containing thimerosal actually increase the risk of autism.[101] One Californian study found that cases of autism still continued to climb quarter by quarter after 2002, when thimerosal was removed from scheduled childhood vaccines.[102]

Meanwhile, other personal and professional charges against Wakefield emerged. Because Wakefield wasn't a pediatrician but a gastroenterologist, he didn't have routine access to a child patient population. Instead, according to a hearing for professional misconduct, he paid children at his son's birthday party ten pounds each to supply blood samples. In the United States, Wakefield's research assistant testified at the 2009 Omnibus Autism Proceeding that the tests he had been instructed to perform, to establish whether there was a connection between bowel disease and autism, were in fact all negative. He claimed in a sworn affidavit that Wakefield was aware of that at the time he and his colleagues published the *Lancet* article, which would have made the paper fraudulent rather than just inadequately evidenced.[103]

Andrew Wakefield was struck off the UK medical register by the General Medical Council (GMC) on May 24, 2010, for serious professional misconduct, including subjecting children to unnecessary and invasive medical procedures such as colonoscopy and lumbar puncture. When the verdict was announced, he immediately launched a public relations campaign from New York, portraying himself to the media as a victim of the British establishment: "This is the way the system deals with dissent."[104] In his 2012 book *Waging War on the Autistic Child*, Wakefield maintains that "the scientific process has been derailed by vested interests acting not on behalf of affected children, but rather for the protection of vaccine manufacturers and incompetent governments."[105] He accuses the chair of the GMC panel of owning undisclosed stock in GlaxoSmithKline and Brian Deer of "traveling aboard Rupert Murdoch's runaway media train." "Deer's story of me, the evil doctor, meant Medicine could be exonerated, public health policy had no hand in the autism catastrophe, and vaccines really are a miracle. Some are still dreaming."[106]

So where does the evidence on childhood vaccination now stand? A literature review performed in 2006 found very few studies showing any association between autism and vaccination, with the vast majority showing no causal link.[107] In November 2010, the American Academy of Pediatrics updated its statement on the evidence base in peer-reviewed medical journals.[108] Among the conclusions from the evidence base were these findings:

1. Despite the mounting total of required vaccinations, children who received the full roster of recommended injections performed better on some neuropsychological outcomes, and no worse on any indicators, than those who had delayed or missed vaccinations.[109]

2. In one study of children vaccinated before a diagnosis of autism, autism risk was actually *lower* in those who had received either MMR vaccine or single-antigen measles vaccine than in nonvaccinated children.[110] Several other studies concluded that the data do not support any association, either positive or negative, between MMR vaccine and autism spectrum disorders.[111] One of the largest of these studies finding no correlation was a *Lancet* article comparing 1,295 young adults with pervasive developmental disorders and 4,469 controls.[112] This study likewise found that MMR vaccination was not associated with an increased risk of such conditions.

3. There is strong evidence against the link posited by Wakefield et al. between autism, persistent measles virus in the gastrointestinal tract, and MMR vaccine exposure.[113] An earlier study in the *British Medical Journal* also found no evidence that children with autism were more likely than a group of controls to have had gastrointestinal disorders at any time before their diagnosis of autism.[114] A fourteen-year-long study of three million reports of severe

gastrointestinal symptoms lasting twenty-four hours or more after MMR vaccination found no data supporting a link to pervasive developmental disorder or inflammatory bowel disease.[115]

But while none of the above studies found any correlation between autism and MMR vaccination, one UK study did find that the proportion of parents blaming MMR vaccination for their child's "regression" to autism had increased since August 1997, when the *Lancet* paper first made the news.[116]

> We found that the trigger for regression in some cases had changed after the publicity and media attention associated with, and preceding, the 1998 Wakefield *et al* paper. Before August 1997, parents incriminated trigger factors such as domestic stress, seizures, or viral illness. Post-1997, parents were more likely to attribute regression to vaccination, especially the MMR vaccine.[117]

THE GLOBAL POLITICS OF THE HPV VACCINE

In contrast to the open public controversy over childhood vaccines, controversy over human papilloma virus (HPV) inoculation has been somewhat less widespread, but it makes up in global reach for whatever it may lack in drama. The theme is the same: have drug company interests corrupted public health decisions and invaded individual autonomy, or has a genuinely beneficial public health initiative been derailed?

In 2006, the Advisory Committee on Immunization Practices of the CDC formally recommended that by the age of twelve, girls should receive a vaccine against HPV, which is linked to cervical cancer. Two such vaccines are manufactured, one (Merck's Gardasil, the first to be licensed) protecting against the four strains of HPV that cause up to 70 percent of cervical cancer in Western countries, and another less expensive vaccine (GlaxoSmithKline's Cervarix) that targets two strains. The vaccines' most immediate function is to prevent the sexual transmission of the virus.[118]

Although it's frequently said that these two vaccines are 100 percent effective against cancer caused by the HPV strains they cover, it hasn't been conclusively demonstrated that they prevent cervical cancer as such—at most, they help prevent the precancerous lesions that can cause cancer.[119] These precursor abnormalities are in fact reduced by almost half in young women who've had the vaccine, according to the first evidence that has emerged, from a nationwide vaccination program in Australia.[120] However, one recent study in the *Annals of Internal Medicine* concluded that "to date, the efficacy of HPV vaccines in preventing cervical cancer has not been demonstrated, while vaccine risks still need to be fully evaluated."[121]

Even if the vaccine were effective, there is additional uncertainty about how long the vaccine's protection lasts[122] and whether women who have received it as girls may be less likely to go for cervical smear tests, because they wrongly assume they're fully protected for life. Additionally, in the huge majority of cases, HPV is an asymptomatic infection that resolves of its own accord.[123]

Yet despite these uncertainties, the general distrust of vaccines, and the culture wars, "the puzzle is that Gardasil prompted comparatively *little* right-wing opposition at the level of federal policymaking."[124] Almost every state has debated whether to make the HPV vaccine mandatory. In 2007, Governor Rick Perry issued an executive order making Texas the first to require that all sixth-grade girls should receive the vaccine. Perry's measure was later overturned by the state legislature as a breach of parental rights, but the fact that it got so far, and that Virginia and the District of Columbia passed and retained mandatory policies, reveals something intriguing about how differently the HPV vaccine has been received by the public. Even while the flames of the MMR fire were still burning, a nationwide poll showed that four in five parents of children under fifteen would give their daughters the HPV vaccine.[125] It's escaped a great deal of the censure that has been heaped on other forms of this "ancient, useless, dangerous, and filthy rite."

But strangely, in parts of the Third World, particularly India, it has incurred much more hostility than other vaccines. Partly that's to do with the high cost of the Gardasil vaccine—approximately $350 per three-dose sequence, ten times the price of the MMR vaccine—and the feeling that the money could be better spent on other public health campaigns or on universal healthcare coverage.[126] Yet as the Botswanan public health doctor Doreen Ramogola-Masire points out, in southern Africa "cervical cancer is now becoming a secondary epidemic in the wake of HIV."[127] Because the HIV virus greatly increases the risk of precancerous lesions, women infected with HIV who are also infected with HPV develop cervical cancer at such an exponential rate that cervical cancer is now the leading cause of death among women in Botswana. Those who work with gynecological cancer patients say that the single most important thing that could be done would be to provide the HPV vaccine to girls.

"When you are sitting in a clinic in Botswana and seeing yet another young woman with end-stage cervical cancer, your realities are different from those you would face in a clinic on the other side of the Atlantic, where your most vexing question is whether a vaccine that potentially protects against a sexually transmitted virus will lead to promiscuity."[128]

Although the incidence of cervical cancer in the United States is far lower than in Botswana, there are issues here, too, about the disproportionate burden of cervical cancer among poor women of color, who are less likely to have Pap smears and follow-up.[129] If Rick Perry's vaccination order for schools had been allowed to stand, it might have had a particularly beneficial effect on Texas's Latina population, who have an incidence rate for cervical cancer of 15.1 percent—half again as high as white women's rate.[130]

While the HPV vaccine was initially thought likely to stir up a firestorm of conservative opposition in the United States on the grounds that it would breed promiscuity, a clever marketing campaign by Merck positioned it as being about preventing cancer. It's hard to oppose a cancer-blocking drug. On September 30, 2005, even before the vaccine was approved by the FDA, Merck had already launched a direct-to-consumer nonbranded promotion called "Make the Connection"—the connection, that is, between HPV and cervical cancer. In 2006, their "One Less" campaign featured active Anglo girls playing soccer and jumping rope double-dutch style, with a Hispanic girl dressing for her rite-of-passage *quinceañera* in the simultaneous "Una Menos" version. All the girls declared "I choose" the vaccine, "because my dreams don't include cervical cancer." Their mothers appeared in the ads too, lending the campaign the imprimatur of maternal approval and distancing it from adolescent sex.

In the "One Less" campaign, boys and men were conspicuous by their absence, although HPV is also linked to penile, anal, and oral cancers.[131] Only two years later, in 2008, did Merck submit data to request FDA approval for marketing Gardasil to young men. Even though it's women who develop cervical cancer, men incur these other cancers—but the message of the "One Less" campaign and the general tenor of the debate relied on the familiar double standard. (The print advertisement for Gardasil ran: "She won't have to tell him she has HPV . . . because she doesn't.")

But the human papilloma virus is spread through sexual contact, and it takes two to tango. If men aren't vaccinated as boys, then heterosexual men are free riders on their female partners' coverage, and gay men aren't protected at all. Yet many commentators present HPV vaccination for men as an altruistic act—with boys being urged to be "chivalrous."[132] That message inaccurately portrays HPV as not being contagious.

Conversely, girls and their parents who elect vaccination are seen as making a Me Medicine consumer choice for their own individual benefit,[133] not as contributing to the public good.[134] Yet herd immunity (usually defined as requiring between 83 to 94 percent of the population) will never be reached against HPV without male cooperation. That's why the rubella (German measles) vaccine is given to boys as well as girls, even though the aim is to prevent abnormalities of the fetus that can occur if a woman contracts rubella during pregnancy.[135] "One of the principal rationales for mandating or encouraging widespread use of the vaccine in girls and boys should be to reduce transmission—that would be to base the campaign for its use on controlling vectorhood—but immunization is actually publicly promoted by emphasizing victimhood instead—and, indeed, the victimhood just of women."[136]

Even for girls alone, however, support for Gardasil began to fall from the initial peak produced by the "One Less" campaign and the vaccine's successful branding as a drug against cancer. Merck was increasingly seen as lobbying for mandatory immunization to clinch its market share before the competing vaccine, Cervarix, was approved.[137] Governor Perry was accused of having ties to the company, which had also given funding to Women in Government, a Washington-based, bipartisan organization of female legislators that had pushed states to add Gardasil to their Medicaid programs and encouraged private health plans to cover the vaccine.[138]

Wags began to quip that HPV meant "Help Pay for Vioxx," Merck's antiinflammatory drug implicated in lawsuits racking up billions of dollars in damages. While in Africa Merck was still seen favorably—the company had donated antiretrovirals and committed $50 million to Botswana's HIV response—public opinion in the United States hardened, with the FDA increasingly seen as having fast-tracked Gardasil under pressure from the firm. As one Colorado parent said, "I guess I just feel like things are rushed through so fast to get on the market without really due diligence done to safety testing. And I feel it's a money thing. You know, they want to get out—go out there and make money. It's a business. And they'll find out later. Wait and see what happens and they've tested us as we go, and I don't feel like that's the right approach."[139]

In India, there has been an even more assertive counterreaction against the HPV vaccine as the creature of Western commercial interests. Because India's HIV rate is not as high as southern Africa's, the way in which precancerous lesions are hastened into full-blown cervical cancer by HPV hasn't been brought out as an argument in favor of vaccination, as it has in Botswana. And while Botswana's progress in keeping HIV-positive people alive on antiretrovirals has led to "technological optimism,"[140] India's headlong rush toward technology-fueled growth, unfettered by government regulation, has left the poorest out and aroused suspicion of foreign multinationals' exploitative motives. As one set of commentators put it, "The case of the HPV vaccine illustrates how promotional practices of drug companies, pressure from powerful international organisations, and the co-operation of India's medical associations to uncritically endorse a vaccine, are influencing public health priorities."[141]

Beginning in July 2009, some 25,000 girls in the Indian states of Andhra Pradesh and Gujarat were injected with Gardasil and Cervarix, supplied by Merck and GlaxoSmithKline, in a project funded by the Bill and Melinda Gates Foundation and administered by the Indian unit of the U.S.-based organization Program for Appropriate Technology in Health (PATH). Most of these ten-to-fourteen-year-old girls were illiterate and poor, many from scheduled castes ("untouchables"). Others lived in boarding schools for indigenous "tribals" under the care of wardens who gave consent on their behalf en masse.

A large proportion of the girls didn't speak the principal local languages, Telugu and Gujarati, and none of them spoke English, the language of the "HPV immunization cards" that they were given. The logo of the government rural health mission was printed on the card, which they probably did understand— as meaning that vaccination was an official policy with which they had to comply. If so, they understood that rightly: it later emerged that the Integrated Tribal Development Authority in Andhra Pradesh had issued orders to dormitory wardens, headmasters, and teachers to comply with the project and "facilitate" the consent-taking process, particularly in the indigenous girls' boarding schools.[142]

As the Indian feminist organization Sama Resource Group for Women and Health discovered in an investigation, there was no infrastructure for reporting adverse events, although it did emerge that seven girls had died after receiving the vaccine. A team from Sama that visited one of the test sites found girls complaining of side effects ranging from heavy bleeding and severe menstrual cramps to headaches and epileptic seizures. But in the absence of proper follow-up, it was impossible to know whether these self-reported complaints stemmed from the vaccine. While there are "concerns that commercial research is being conducted in the garb of providing a public service,"[143] you'd think the companies would want to know the results of the "research."

The absence of follow-up leads to suspicions that the main motive was a marketing opportunity assisted by charitable funding.[144] Here was a chance to forge a public-private partnership with a neoliberal country that has a huge population of potential users for this and other pharmaceuticals.

But in April 2010, the Indian ministry of health and family welfare called a halt to the HPV vaccine trials, in response to expressions of concern from no fewer than sixty-eight campaigning groups. Complaints began to emerge in October 2009—just a few months after the study began.[145] Although GlaxoSmithKline, Merck, and PATH had issued defenses of the process and of the vaccine itself in February 2010, the government announced an inquiry, which produced an interim report in November of the same year. As of May 2012, the final report had been written but not yet released into the public domain.

The interim report indicated no clear connection between the vaccine and what by then was known to be eight deaths—although the committee noted that there was insufficient information about how three of the girls died. It did castigate the researchers for failing to obtain informed consent or to set up a system for reporting adverse events. The researchers were also criticized for concentrating on underprivileged tribals and scheduled caste members as well as for using state public health officers to seek consent, suggesting that this was a mandatory program. Finally, the report concluded that the companies should never have been allowed to evaluate the feasibility of the program themselves, when they had an obvious interest in finding that it had succeeded. But in the end, the interim committee report still wrote all these failings off as "minor."

No health insurance or ongoing care was provided to the girls—although PATH took out insurance coverage for itself—even though the study was done in rural areas with a poor quality of medical care. The researchers claimed that no coverage was required because this was only a "demonstration project," not a study as such—but more Indian girls were vaccinated in this one project than the *total* of 21,000 American girls in the studies that had led to Gardasil being licensed by the FDA.

Suddenly, the research had also now jumped from being a "demonstration project" to being described as a "phase IV post-marketing trial to evaluate strategies for delivery of the vaccine and its acceptance by the population."[146] All clinical trials up to phase 3 have to be registered with the Indian authorities, but not phase 4 studies. So what happened to the previous three phases, particularly the crucial phase 1, which is meant to discover whether a new drug has any adverse side effects? When Sama asked that question, using the

Indian Right to Information Act, it was rebuffed on grounds of commercial confidentiality. But it appears from the trials listed in the U.S. government clinical trial registry that marketing approval for the vaccines in India was based on two very much smaller phase 3 trials on a total of just 450 girls, as well as on FDA approval.[147]

It seems abundantly obvious that the process was flawed and deeply undemocratic—but if the study had been handled better, might HPV vaccination have been of some benefit to the girls? Critics have cited the government's acknowledged duty not to experiment on indigenous peoples unless they stand to benefit. The very absence of cervical screening and decent health services might well mean that the vaccine could benefit rural women of the scheduled castes and tribal groups, particularly in light of the Australian findings that precancerous lesions are halved in those receiving the vaccine.

Another Australian study has found a higher prevalence of some HPV strains among indigenous women than among other women in the same thirty-one-to-forty age group, 35 percent as against 22.5 percent. They were also more likely to smoke, a risk factor, and to have Pap test abnormalities. Despite a national cervical screening program, which India lacks, "indigenous women in Australia have a disproportionate burden of cervical cancer."[148] As in Botswana, might there be an argument that the most disadvantaged groups could be helped by HPV vaccination if it were effective?

Yet it's not at all clear that the four strains of HPV covered by Gardasil or the two strains targeted by Cervarix are the dominant pathogens in India: in some countries in Africa, such as Gambia and Zambia, completely different strains are the cancer-causing ones.[149] And when coverage for basic and much cheaper childhood vaccines such as measles is declining countrywide,[150] is an HPV vaccine the real priority for India?

Whatever the merits of the vaccine itself, it's evident that the Indian antivaccination campaign wasn't just peopled by crank conspiracy theorists—although it was accused in a *Lancet* article of presenting "a mix of facts, misinformation, principles, anger and injustices, and anger at perceived interference by multinational companies, funders and international non-governmental organisations in local concerns."[151] A response from Sama activists and other Indian experts reiterated that vaccines marketed by First World companies in Third World countries need to meet genuine Third World needs.[152] Other vaccines that would count as routine childhood immunizations in the United States may not qualify.

But just as there has been "vaccine creep" in the United States toward increasing numbers of injections, "there has been pressure from the drug industry to include all newly developed vaccines in the [Indian] government's UIP [Universal Programme of Immunization], even though the clinical and epidemiological justification for their inclusion is debated."[153] A recent official report from a committee of the Indian parliament strongly condemned the Indian regulatory body, the Central Drugs Standard Control Organisation (CDSCO), for putting the profits of the drug industry above the needs of the consumer:

> For decades it [CDSCO] has been according primacy to the propagation and facilitation of the drugs industry, due to which, unfortunately, the interest of the biggest stakeholder i.e. the consumer has never been ensured. Taking strong exception to this continued neglect of the poor and hapless patient, the Committee recommends that the Mission Statement of CDCSO be formulated forthwith to convey in very unambiguous terms that the organization is solely meant for public health.[154]

Although vaccines are the epitome of We Medicine, in a system where only the rapidly expanding middle class can afford good health care, new vaccines are presented in India as must-have private purchases in a Me Medicine manner. "Aggressive promotional campaigns for the new vaccines, and their quick adoption by industry-friendly private medical practitioners, have already made these vaccines akin to fast-moving consumer goods."[155] These vaccines are typically highly priced, like the HPV vaccine, and they account for most of the rapid growth (8 to 10 percent per year) of India's vaccine market. Traditional, less expensive vaccines are being phased out: between 1998 and 2001, ten out of fourteen major manufacturers stopped making them.[156] As with flu vaccines, there was little or no profit to be had.

Ironically, however, this development isn't part and parcel of personalized medicine but its opposite: "For every new drug invented, genomics and bioinformatics are used to customise it to suit different populations. But in vaccines, the tendency is to move toward a 'one vaccine fits all' regime. Though there were no attempts to conclusively establish that the imported vaccines actually suited the Indian strains of the pathogens, these vaccines were adopted."[157]

We saw in chapter 3 that pharmacogenomics may push U.S. drug manufacturers into niche marketing, stratifying patient groups and moving away from mass-market blockbusters. In the Third World, the opposite dynamic seems to be at work: one-size-fits-all vaccines are marketed that may not match the local strains of pathogens or meet the most urgent local needs. These Third World markets can continue to be "off the rack" rather than "tailor made." That would give us the worst of We Medicine—its lack of specificity—along with the worst of Me Medicine, its consumerist outlook.

THE FOUR HYPOTHESES

Can the four hypotheses explain why vaccination has been portrayed as a public enemy? Do the same factors that account for the widespread acceptance of Me Medicine also underpin popular hostility to vaccination as a form of We Medicine?

Threat and contamination are recurring motifs in the long history of fear and loathing surrounding vaccination as a "dangerous and filthy rite." In chapter 1, I pointed out the way in which contamination and pollution, as powerful motivating fears, can extend to many forms of "dirt" and impurity.[158] In popular accounts of the time, vaccination's origins in cowpox inoculation breached the taboo against interspecies contamination.

When Jenner first introduced his technique, it was welcomed as a huge improvement on the more dangerous method of variolation, in which virus procured directly from a pustule of a sick patient was scratched into the skin. But there were also hostile comments that Jenner was engaging in "the injection of the morbific matter of a diseased animal into a healthy child."[159] The satirical cartoonist James Gillray produced an etching showing Jenner's previous patients mutating into cows, while the good doctor applies his lancet to the arm of another victim.

Because vaccination is performed on healthy patients, it's uniquely liable to these fears, as there is no countervailing immediate benefit to be balanced against the risk of contamination. Some early doctors actually saw vaccination as a violation of the Hippocratic oath, breaching the injunction to "first do no harm." When the healthy patients to be vaccinated are children, the sense of threat is all the more overpowering. Yet in another sense, the problem is that when vaccines work well, there's not *enough* sense of threat: people easily forget that measles and other childhood diseases still kill.

It's not just perceptions at issue. Contamination has sometimes been a literal threat: for example, during the 1901 U.S. smallpox epidemic, over ninety people died after the use of contaminated batches of vaccine and unsterilized instruments. In other cases, however, contamination was genuinely but wrongly feared, as in the statement by the Nigerian Muslim group that polio vaccines were "corrupted and tainted by evildoers from America and their Western allies." In Nigeria, the threat from the disease targeted by the vaccine, on the other hand, *was* genuine: the country accounted for 45 percent of all polio cases worldwide and 80 percent of reported cases in Africa. But suspicion of public health officers was still strong enough to overcome the perceived benefits of vaccination and to dash the campaign's hopes of immunizing more than fifteen million children throughout western and central Africa. Where people struggle to obtain more immediately urgent medicines, "to be offered free medicine [inoculation] is about as unusual as a stranger's going door to door in America and handing over $100 bills."[160]

In wealthier countries, where the threat of serious contagious disease is perceived as historical—it's been nearly a century since the Spanish flu epidemic—dread of contamination also readily leads to the judgment that the vaccine is more to be feared than the disease itself. But it would take just a few mutations in the H1N1 virus—genetic tweaks whose course scientists now understand in animal models—for pandemics to kill millions worldwide. Yet despite this threat, vaccines are widely seen as just another toxic substance,[161] like the 320 toxins for which David Ewing Duncan had himself tested. The thesis behind Duncan's *Experimental Man* was that the way to protect yourself against the environmental hazards inflicted by even the most apparently idyllic American childhood was through personal genetics.

Me Medicine is the solution, in this view, while We Medicine in the form of vaccines is actually part of the problem. That helps explain why vaccination programs meet with apathy or outright hostility. In a minority of instances, however, the threat from toxic side effects of vaccines has also been genuine—for example, in instances of contaminated vaccines or unconsented "trials" such as the Indian HPV case.

What about the second explanation, narcissism? It would be offensive and crass to brand the concerned parents who flocked to support Andrew Wakefield mere narcissists. As elsewhere in this book, particularly in the case of private cord blood banking, what looked like the most obvious cause of Me Medicine's popularity turns out to be the weakest. In the case of vaccine denialists, however, the related "bowling alone" idea has some merit: it can be applied to the "I know better than the experts" mentality actively encouraged by the media in the MMR controversy.

When the model and TV personality Jenny McCarthy's son Evan was diagnosed with autism, she became convinced that what she called "the autism shot" was responsible. On the *Oprah Winfrey* show in 2007, she explained that "Mommy instinct" lay behind that conviction, whatever the doctors said. Against a CDC statement on the program that "the vast majority of science to date does not support an association between thimerosal in vaccines and autism, [and] it is important to remember: vaccines protect and save lives," she replied, "My science is named Evan. . . . The University of Google is where I got my degree from."

Although McCarthy later distanced herself from her earlier position that vaccines directly caused autism, expressing a desire for further research, her message on the *Oprah* show is an obvious instance of rejecting the expert consensus—of bowling alone. Andrew Wakefield also portrayed himself as a solitary rebel against the medical consensus when he retorted to the GMC tribunal that had stripped him of his license to practice, "This is the way the system deals with dissent." But a society of solitary bowlers is a public health disaster waiting to happen.

> This "I feel, therefore it is" outlook may empower Winfrey's viewers to take charge of their lives, but it ignores completely the perils that face a society where everyone runs around intuiting their own versions of the truth. The notion that people should base medical decisions on what is "right for them" is particularly problematic in a public health context, where individual choices cannot be cordoned off from each other.[162]

Willingness to be vaccinated can be seen as a form of social capital, which overcomes the free-rider problem—exemplified by that self-contradictory statement, "Since most children are vaccinated, our nation has enough 'herd immunity.'" If individuals feel that they have a stake in *everyone*'s health, then they won't try to excuse their own failure to vaccinate their children by letting other parents take the strain. Population immunity is itself a kind of cooperative, a form of collective wealth in which we all share and from which all of our children benefit.

But it depends on the scale of the collectivity in which you feel you have that social capital and the purpose to which it's put. When Jackson, Tennessee, residents installed armed guards on trains into their town to protect themselves against strangers with smallpox, detaining them forcibly in impromptu prison camps, they were investing their social capital at a local level but also disregarding basic human rights. Antivaccine groups in the MMR scare also invested their social capital in parents and children in a self-identified group, with the aim of discouraging other parents away from vaccinations—with the result that children below the age of vaccination and the elderly may have been put at increased risk from measles outbreaks.

In these two examples, people weren't "bowling alone," it's true—but social capital can be exclusive as well as inclusive. Back in chapter 1, I made a similar point about civic associations more generally: "engagement in community affairs and the sense of shared identity and reciprocity," for which Putnam lauds the postwar period, included support for organizations deliberately dedicated to excluding others, from the Masons to the Daughters of the American Revolution. But the example of vaccines does prove Putnam right in another sense. We've seen that the high tide of popular acceptance of vaccination coincided with what Putnam identifies as the peak of civic togetherness in America. Since the Salk vaccine for polio was developed in the 1950s, however, there's been no consistent and wholehearted defense of vaccination and public health in the United States: rather, all too often, alternating apathy and hostility.

Corporate interests: As with threat and contamination, it's certainly true that there's a widespread popular perception at work here. Vaccine resisters worldwide have accused pharmaceutical companies of continually inflating the number of inoculations recommended or required, with government regulatory agencies allegedly in their pockets. Whereas we hear little about corporate interests in bright and shiny new areas of Me Medicine such as direct-to-consumer genetic testing, it's routine for antivaccinationists to portray drug companies as the *eminences grises* behind officially recommended vaccination policies. Even the potentially vast threat of pandemics is sometimes dismissed as a fabrication by drug companies to sell their products—even though during the swine flu pandemic the pharmaceutical companies were unable to meet the existing demand for their vaccines. (Antivirals are a separate issue.)

Vaccine resistance embodies rejection of the notion "public good," but it's skeptical about private interests—so that, unusually, the message is "public bad, private bad too." But in the case of threat and contamination, the perception, while powerful, was mainly incorrect (except for isolated contamination instances). Is the widespread belief that drug companies dictate official vaccination policies likewise just a misperception?

This question is complicated by the differentiation of markets in vaccines between the rich and poor countries as well as within countries. Lesser corporate interest in cheaper, low-return or high-labor-input mass vaccines, such as those for measles or influenza, makes it less plausible that drug companies controlled public health responses in the swine flu case in order to bump up demand. According to the bioethicist Art Caplan, vaccine production accounts for only a small proportion of what the drug industry does and possibly even less of what it really cares about.

Low returns, plus unpredictable demand, plus mutating viral strains, plus liability suits all add up to an unattractive business plan. Eli Lilly, Dow, Pfizer, Cutter, and Parke-Davis have simply stopped making vaccines. Although some of the decline is down to corporate mergers, it's still surprising that there were twenty-six U.S. vaccine producers in 1967 but only four companies making vaccines fifty years later. "Biologicals"—such as vaccines, made by modifying a disease virus or bacterium, as opposed to chemically based mass-market drugs—"barely qualified as footnotes in the official histories of these firms."[163]

Since the turn of the present century, there have been serious shortages of influenza, MMR, DTP (diphtheria, tetanus, and pertussis), chickenpox, and pneumococcal vaccines.[164] But at other times—as during the swine flu pandemic—huge numbers of doses had to be thrown away. True, government health services may underwrite those losses: roughly 55 percent of childhood vaccines are now bought by the federal government under the Vaccines for Children program. But vaccine manufacture has faced a profit crisis since the 1970s, when *Science* speculated on the forthcoming demise of the industry in an editorial headed "Our Last Vaccine?"[165]

Traditional vaccines "were square pegs that didn't fit into the triangular holes of market capitalism."[166] Newer high-return products such as the HPV vaccine are a different matter, both in terms of profitability and in the resulting interest by firms in persuading governments to let them have access to their populations. The Indian example does seem to demonstrate corporate capture of public health authorities, with inadequate consent procedures, poor research ethics, and lack of transparency. Where corporations can reap favorable public relations through involvement with a major charity like the Gates Foundation, however, they may be willing to take a loss or even to donate the vaccine—as in Botswana.

In the HPV case, we can also see another phenomenon that recurs throughout this book: capitalism's talent for creating new products and markets where none existed before. That also applies to the MMR backlash: Big Pharma, in cahoots with government, was accused of continually upping the ante for new childhood vaccinations. What's intriguing here is that "new" carries unfavorable connotations: that's something . . . well, new.

With enhancement at the extreme, Me Medicine rides on the crest of popular enchantment with new technologies. But when vaccine developers shake off the "ancient and useless" tag with innovative therapies, they meet with distrust. While the methods that manufacturers have used to gain new markets certainly deserve suspicion, as in the Indian case, the HPV vaccine itself may be of genuine benefit in Third World countries where cervical cancer rates are high but screening and treatment nonexistent, as in Botswana. On the other hand, critics allege that the drug industry has put the Indian government under pressure to include new vaccines in universal immunization programs just because they *are* new, even if they don't meet particular national needs.[167]

Neoliberal policies in India have already been mentioned as creating a situation in which official regulation and oversight were weak—although government authority was also used to promote acceptance without proper informed consent among a vulnerable group. Government's role, in this unattractive extreme of neoliberalism, could be seen as carrying a big stick for individuals but offering a juicy carrot for corporations. Less blatant but recognizably neoliberal policies are also relevant to the United States, with budget cuts and outsourcing crippling the response of public agencies to outbreaks of infectious disease—unless the watchword "terrorism" can be applied. After 9/11, Congress authorized $1.9 billion to stock smallpox vaccine and $1.4 billion to build an anthrax vaccine storehouse, spending a total of $33 *billion* between 2002 and 2006 on "biodefense."[168] Meanwhile, in 2003, the NIH invested less than $70 *million* on influenza vaccine research in 2003. Extrapolated over the same five-year period, that level of expenditure would come to just a bit over one-thousandth of congressional spending on "biodefense."

In chapter 2 we saw that President Obama's budget for 2012 committed the federal government to a 2 percent overall cut for the Department of Health and Human Services but a 2.4 percent increase for the National Institutes of Health. Most of that money will go not to mass vaccine development but on personalized medicine—what Francis Collins of the NIH called "leveraging new genomics technologies in disease and health research and translational science, and in pursuing goals in personalized medicine."[169] This concerted support from public agencies such as the NIH is exactly what the personalized medicine industry wants to see, as is clear in this classically neoliberal extract from a keynote speech in May 2012 at the Eighth Annual State of Personalized Medicine Luncheon, given by the industry's Personalized Medicine Coalition:

> There remains great opportunity for the further personalization of health care; but business and venture capital must partner with policymakers to create an ecosystem that rewards innovation and recognizes the opportunity for personalized medicine to transform the health care system. Regulatory and reimbursement policy must catch up with innovation in order to bring new treatments to patients. Business incentives must be put in place to encourage private investment and further develop the pipeline of new personalized medicine products.[170]

A great strength of vaccination programs, in the eyes of their advocates, is that they mobilize all the state's resources for all the people, not only for business and venture capital. As Bill Foege, the former director of the CDC, puts it:

> From my perspective, immunization is a foundation stone for all modern public health because it's inexpensive and democratic in that it provides social justice for everyone. . . . It involves every aspect of public health. You have to do surveillance, you have logistics problems, you need teams of people and evaluation. And so it becomes training wheels for other public health. And if you can't run a vaccine program it's not likely you can do other things.[171]

But to committed neoliberals, the idea that if you can run a vaccine program, you can do other things is a threat, not a promise. To them, those strengths are just exactly what's wrong with vaccination programs: they multiply government interference with individual freedom.

Choice and autonomy are certainly the rallying cries of vaccine skeptics, but how accurate is that explanation? Is it simply the case that vaccination as a form of We Medicine is unpopular because it restricts the freedom of individual parents and patients? Does that only apply to mandatory vaccination? Powerful players freely use both kinds of argument—public benefit and private choice—when it suits them. If it's true that Merck lobbied for mandatory Gardasil immunization, that sits rather oddly with the "I choose" language of the "One Less" campaign.

In the MMR case, the congressional hearings and media coverage seemed to be not so much about offering parents an informed choice as about telling them definitely *not* to choose the vaccine. And then there were the death threats, like the one received by the Institute of Medicine committee on vaccine safety: "Forgiveness is between [the committee's members] and God. It is my job to arrange a meeting."[172]

Even when choice really does seem to be a genuinely important value for vaccine skeptics, it's often masking another issue—class in the English revolt against compulsory smallpox vaccination, race and ethnicity in the U.S. context, or anticolonialism in Indian activists' resentment of Western drug company incursions. We need to understand these profoundly social factors to locate modern vaccine resistance in the historical and political context. Individual choice is only a superficial explanation.

Nor does the debate about vaccination turn on communal good versus individual choice—not if all of us are both victims and vectors. But it's quite true that this is how it's usually perceived. In turn, popular resentments and the alleged failures of vaccination programs are mustered as evidence against the whole idea of collective benefit. In the last chapter, I want to ask how, in the face of such continued assaults, we can reclaim biotechnology for the common good.

7

RECLAIMING BIOTECHNOLOGY FOR THE COMMON GOOD

GIVING PERSONALIZED MEDICINE AN OBJECTIVE REALITY CHECK has revealed that the evidence base is uneven. In pharmacogenomics, there have been some genuine advances, but in enhancement and retail genetics evidence for the supposed revolution is largely lacking, while in private umbilical cord blood banking the medical consensus is that the procedure may actually be dangerous. Whatever explains the rise of Me Medicine, it isn't just the science behind it. Likewise, the causes for popular revulsion against vaccines as the most prominent form of We Medicine aren't rooted primarily in medical evidence, most of which is rejected by many vaccine denialists.

So if the scientific evidence alone doesn't explain the rise of Me Medicine and the comparative decline of We Medicine, what does? In chapter 1, I introduced four possible hypotheses with wider social and political explanatory power. Of the four hypotheses, the explanation in terms of *corporate interests and neoliberal public policy* has proved the most consistently robust, although less so in the case of vaccination—the one area where popular opinion is very quick to claim that it's all down to hard cash.

In all the Me Medicine case examples, the public sector, as the entrepreneurial state, is being asked to sponsor the growth and shoulder the risks for the private sector. Private capital has not simply relied passively on a permis-

sive government regime in areas such as genetic patenting but has also actively sought state backing in rolling back regulatory measures and in developing the blue-sky science that can then be "translated" into profit-making
services such as retail genetics. These demands are exemplified in the words of
the keynote speaker at the personalized medicine industry's 2012 annual luncheon: "Regulatory and reimbursement policy must catch up with innovation
in order to bring new treatments to patients. Business incentives must be put
in place to encourage private investment and further develop the pipeline of
new personalized medicine products."[1]

Personalized medicine has created risks as well as rewards for firms. For example, drug companies accustomed to relying on mass-market blockbusters
will have to develop a different business plan if pharmacogenetics is to become
commercially viable. But Me Medicine illustrates capitalism's flexible talent for
creating new products and markets where none existed before. That, rather than
scientific plausibility, goes a long way toward explaining private umbilical cord
blood banking and retail genetics. Undermining the distinction between treatment and "cosmetic neurology" or other forms of enhancement also creates new
product possibilities, by making the quest for self-improvement potentially
limitless.

The relevance of *threat* is most obvious in the pharmacogenetics and vaccination examples, although it is also significant in retail genetics and private
umbilical cord blood banking—but in both the latter cases, *promise* is more
prominent than threat. Inflated promise is certainly predominant in enhancement as well, although threat is relevant to military uses of enhancement technologies. In pharmacogenetics, the dual threat of cancer and iatrogenic harms
from oncology treatment does a good job of explaining the appeal of personalized drug regimes to patients, although not the scientific basis in molecular pathology for the development of pharmacogenetics. In the vaccination case, too,
threat was a powerful theme, but it was Janus faced: while the substantial threat
from pandemic disease often met with public apathy, the side effects of vaccination against such diseases seemed an even greater threat to those who opposed
the MMR vaccine.

Choice and autonomy are crucial to the rhetoric of Me Medicine and a
powerful rallying cry against We Medicine in the form of vaccination. Vaccine manufacturers have fought fire with fire by using the language of choice
to promote their products, for example the "I choose" language of Merck's
"One Less" campaign. However, in retail genetics the rhetoric of choice, autonomy, and self-ownership is at odds with the reality of who actually owns

property in genetic samples and databanks. In pharmacogenetics, choice does seem to be enhanced in a few areas, such as the option of lower hormone dosages in IVF, but it's not at all certain that patient choice really will be the name of the game overall. That will depend on the way pharmaceutical companies and insurers deal with the economic consequences of patient stratification into smaller niche markets. Nor was choice a prominent theme in private cord blood banks' advertising: duty and responsibility appeared to be more powerful motivators. The same is true of enhancement, likewise presented by its advocates as a moral must. Additionally, enhancement may actually diminish choice by leading to a race to the bottom for those who can't afford the necessary drugs or neuroenhancement technologies.

Narcissism, which might have appeared the obvious frontrunner among the four hypotheses, wound up in last place. Even though it might seem particularly applicable to enhancement, advocates of those technologies instead stress beneficence and duty, for example, to produce the best children we can. While retail genetics does sometimes seem to play on the customer's narcissistic sense of being uniquely special, one third of customers genuinely believe they're buying a diagnosis, which isn't narcissistic as such. It seems offensive to regard a mother's willingness to undergo the further procedure needed to collect umbilical cord blood for private banking as in any way narcissistic, and the same applies in the case of cancer patients attracted to pharmacogenetic therapies. The related "*bowling alone*" thesis fared slightly better in explaining popular distrust of expert evidence about vaccination, but it ran into difficulty in other cases. For example, it couldn't explain the large number of public blood banks, a form of social capital.

Given that the scientific evidence doesn't dictate that "you have to be ready to embrace this new world" of personalized medicine, as Francis Collins urged, where do we go from here? Is Me Medicine or We Medicine the way forward? Must we choose? The answer to this question, again, can't just depend on the scientific evidence, although it is important to know what that evidence is before we can give an informed answer. That's what I've tried to do in the previous chapters, drawing together an up-to-date evidence base across the entire range of personalized therapies. Now that I've shown how much we know about the science and what nonscientific factors such as corporate interests are involved, the next step is to look at the political landscape in which Me Medicine and We Medicine have to exist.

The contours of that countryside are strongly shaped by a strange phenomenon first encountered in the vaccine case: *communal* solidarity in the cause of absolute *individualism*. Parents' groups opposing the MMR vaccine developed powerful networking and communication strategies but preached a message of individual right to refuse—a form of free riding that would have destroyed population immunity if followed by every family. These groups seem to distrust big government and big business equally, seeing public health as a public enemy rather than as a source of protection not only against microbes but against pharmaceutical companies' interests. Instead, many of these groups blamed corporations and government pretty much equally.

You could find this phenomenon deeply frustrating and dismiss the vaccine denialists as crackpot conspiracy theorists. I wouldn't blame you if you did, particularly when the wilder fringes of the movement engaged in death threats of the sort Paul Offit received. But you could also find evidence in the antivaccine movement of a kind of communal coming together. It's certainly not what Putnam had in mind when he called for us to stop bowling alone, but it added up to an impressive popular movement. These parents were building on what they had in common—concern for their children—even if the message was the individual parent's absolute right to refuse.

Similarly, in the American politics of the early twenty-first century, the renaissance of the Right—as manifested in what President Obama called the "shellacking" taken by the Democrats in the midterm congressional elections of 2010—has used the power of common concerns for a paradoxical purpose. They've employed communal action to present their message that communal action is pernicious and to elevate economic individualism to the status of the sole value. What happened in the vaccine instance foretold what would occur later in national politics: people came together as We to demand that only Me should count.

In the rest of this chapter, I want to ask whether we can use communal action for its rightful communal purpose instead—to reclaim biotechnology for the common good. I'll begin with an astute recent analysis of why communal action groups at the national political level—not just in healthcare—have used the power of the many to support the privileges of the few. I'll then move on to the venerable notion of the commons and show how it's been resurrected in an unexpected and largely unknown corporate form. After examining the way in which other sorts of patient groups have challenged that commodification of the commons, I'll end with an optimistic account of how we can work together for what we have in common, to keep the notion of We Medicine alive and well.

HOW "WE THE PEOPLE" BECAME "WE THE MARKET"

The political analyst Thomas Frank has recently studied, at a national level, the same sort of communal awareness in the cause of fervent individualism that arose in the vaccine example. In his 2012 book *Pity the Billionaire*, Frank asks why the economic meltdown of 2008 didn't produce the sense of collectivism that the United States witnessed after the stock market crash of 1929. The assumption, he thinks, was that it would: that the We factors of solidarity and altruism would motivate not just government policy but also our everyday transactions with each other. "As the economy falls apart, the assumption goes, we will also rediscover a certain neighborliness, a sense of community and even collectivism that comes from shared privation . . . [just as] the 'rugged individualism' extolled by Herbert Hoover gave way to the generosity and social solidarity documented by Studs Terkel in his oral history *Hard Times*."[2]

That supposition was particularly plausible because the economic crisis stemmed from lack of supervision of financial institutions, which was part and parcel of "the familiar neoliberal faith that market forces would cause financial institutions to regulate themselves."[3] Those policies included exempting some derivatives from regulatory oversight, allowing banks to bypass previous requirements about a certain level of cash holdings, overruling state predatory lending laws and permitting investment banks to amass and sell portfolios of subprime mortgages without the consent of the homeowners. You might have thought that there would have been a shared realization that speculative banking and loose regulation had created the disaster for which we were all now bearing the burden.

What happened instead? Unexpectedly, successful populist movements grew from the bottom up—such as the Tea Party, built on local chapters and mass rallies, or Freedom Works, the free-market pressure group whose activists rely on the methods of the 1960s neighborhood organizer Saul Alinsky. (True, the Tea Party has friends and funding in high places, but it prides itself on its grassroots style of organization.) If the form of these movements was communitarian, however, their message wasn't: they do blame Big Business for the economic debacle, but they blame Big Government as much or even more. Just like the parents' resistance movement in the vaccination case, these groups see the corporations and the state as acting hand in glove. They don't view regulation as a bulwark against further banking crises, nor do they have faith that the government can turn the economy around. Instead, they want to give the market even freer rein, in the conviction that its ailments will eventually cure themselves when purged by the tonic of recession.

Because many Tea Party supporters are small businessmen and storekeepers who have to meet their balance sheets or go bust, they saw the national deficit as no different from their own situation. Lumping together Lehman Brothers and the overstretched Las Vegas homeowners who'd leveraged themselves into unaffordable subprime mortgages, they accepted the free-market argument that the nation had to be cleaned from corruption and "quantitative easing," as well as from crony capitalism and socialism—seeing both pairs as much the same. "Regulation, too, was merely a conspiracy of the big guys against the little."[4]

> The market's invisible hand would lift the threat of "destruction" from the land. It would restore fairness to a nation laid waste by cronyism and bailouts. It would let the failures fail, at the same time comforting the thrifty and the diligent. Under its benevolent gaze, rewards would be proportionate to effort; the lazy and the deceiving would be turned away empty-handed, and once again would justice and stability would prevail.[5]

How is this relevant to Me Medicine? Market populism could be at work there too. Perceiving the U.S. healthcare system as having failed might well lead you to the view that you've got to protect your own health, for which you're individually responsible. The support you need would then come not from socialized medicine or even Obamacare but from your own personalized healthcare package. That's certainly the message put across loud and clear by retail genetics.

Just as the renewed Right sees the problem with the free market as being that there isn't enough of it, even in the highly individualized American medical system advocates of personalized medicine claim that healthcare isn't individualized enough. In some instances, as with pharmacogenomics, there are some good scientific arguments behind that claim, but other forms of Me Medicine such as private cord blood banking seem to benefit from an automatic assumption that we badly need more individualization, whether or not that's backed up by evidence.

Frank argues that in the economy more generally, "We the People" has degenerated into "We the Market." That chimes with Michael Sandel's analysis: rather than *having* a market system, he says, now we simply *are* a market system, as if it possessed us rather than the other way around.[6] The free market is classically seen, however, as no more than the sum of all our individual choices, not as a social institution with a mind of its own. It's We, all right, but only in the most minimal sense: it's really no more than all our Me's put together.

Similarly, when the very notion of public health is distrusted and the notion is promulgated that I'm solely responsible for my individual health, Me Medicine comes to be seen as all the We that's necessary to deliver us from the threats hanging over healthcare, just as its equivalent in economic policy is seen as lifting the threat of financial meltdown. With the additional glamour of new biotechnologies behind it, Me Medicine thus appears to its advocates as the inevitable and desirable way of the future.

"Utopian capitalism can look pretty good in a time of disillusionment and collapse. It is a doctrine that seems to have all the answers."[7] So, too, does personalized medicine, particularly when presented in a utopian manner as "a true revolution."[8] Behind that standard, the troops are massing, but in at least one particular case, critics have suggested that they're really just cannon fodder. This example of a private corporate biobank created by individual paying customers shows just how complex and counterintuitive the economics and politics of personalized medicine have now become.

HOW "WE THE RESEARCH PARTNERS" BECAME "WE THE PRIVATE BIOBANK"

> We believe research is a two-way process, where participants are valued as partners in scientific discovery.
> —23andMe blog "The Spittoon," June 16, 2011

The retail genetics firm 23andMe, first described in chapter 2, also has a research arm called 23andWe. When they log in to their accounts after receiving their initial results, customers are invited to fill out online surveys about their lifestyle and health, which add value to the genetic data and create a sizeable biobank. About 60 percent of 23andMe customers have agreed to provide this information, resulting in a database that already numbered over one hundred thousand individuals as of June 2011.

Green or blue dots near customers' avatars proclaim them to be "research pioneers" or "research trailblazers," depending on how much data they've contributed. They can even become "research captains" by recruiting other customers. While the firm's website refers to customers who provide both spit and surveys as "collaborators," "advisers," and "contributors," elsewhere CEO Anne Wojcicki calls them "active genomes."[9] The company prides itself on what it calls its "participant-led" paradigm, distinguishing it from conventional research.

Of course, it's also distinguished from conventional research in another way: research participants generally either volunteer or get paid—rather than pay—to take part. With the exception of programs such as "Roots Into the Future," in which the firm has offered free genetic tests to ten thousand African Americans, and free kits for people with particular genetic conditions including Parkinson's disease or the rare blood and marrow cancers called myeloproliferative neoplasms, 23andMe customers are just that: customers, not research subjects. Their tissue and the data they willingly provide does constitute the raw material for research, but that doesn't make them equal participants in scientific progress.

While research is "a two-way process," however, profit making appears to be a one-way street. There's no element of communal benefit sharing, even though that notion became official policy in genetic research over twelve years ago, when the ethics committee of the Human Genome Organization at the WHO recommended that between 1 to 3 percent of profits should be returned to participants.[10] It wouldn't be beyond the considerable wit of 23andMe's strategists to offer those "research pioneers," "trailblazers," and "captains" a contribution to a medical charity of their choice rather than the equivalent of a gold star on your homework. While some commentators worry that benefit sharing might constitute an undue inducement to participate in research,[11] obviously that argument doesn't apply where participants already pay to participate.

As part of this overall reality check for personalized medicine, it's necessary to bear in mind that 23andWe participants are paying twice over—with hard cash and with the time that they put into answering the lifestyle survey questions that yield the epidemiological data. But that's not my main point. Instead, I want to concentrate on the way in which a valuable communal biobank, with additional potential profits from patenting, is being built up from all of these individual participation efforts. Rather than being public, however, that database remains a private commercial resource.

It's not unknown or impossible for large-scale genetic biobanks to be more communal in character. UK Biobank (UKBB), five times the size of 23andMe's biobank, is constituted legally as a corporation. Its contributors don't own either the samples they've provided or the additional epidemiological data they supply. But the biobank's ownership of its samples is constrained by a legal obligation as a nonprofit entity to act consistently with its charitable purpose, by the fiduciary duties of its directors, and by their legal position as charitable trustees as well as company directors under UK corporate law.[12] Its "Ethics and Governance Framework" states that UKBB won't exercise all the possible rights of ownership: it won't sell samples outright, for example. Donors also have the right to withdraw their samples, although that wasn't originally mooted because such unpredictability was thought to undermine the scientific and commercial value of the collection.

If 23andMe were serious about partnership, it could reconstitute 23andWe as a not-for-profit arm and emulate UK Biobank's arrangements. Those provisions are by no means perfect, but at least they incorporate some attempt at community benefit beyond the rhetorical level. During the consultation in 2002 leading up to UKBB's establishment, the British public expressed concern that the biobank would focus on "profitable diseases" rather than public health. It does seem that some of their We-focused worries have been taken on board, as is appropriate for a national resource derived with public funding and NHS patients' crucial contributions—even though the UK Biobank model still falls short of full communal ownership.

Another possible model is joint patenting between patients and researchers, pioneered by an advocacy organization, the PXE International Foundation, which was set up by families of children with the connective tissue disorder *Pseudoxanthoma elasticum*. Unlike the legal situation with either UK Biobank or 23andWe, the PXE families really are joint owners of the biobank, but they don't take profits: those go back into further research.[13] Here's an example of a new form of community, genetically defined, acting on behalf of a collective rather than according to the principle of individual choice. Patient advocacy groups have existed for a long time, of course, but this is something different. "PXE International has become a well-known model for the way it has leveraged its control of the biobank *qua* biocapital in order to achieve collective goals."[14]

Both these precedents suggest a possible middle way between pure altruism and pure capitalism.[15] These models make use of the market, but, particularly in the PXE example, they attempt to put right "a market failure with respect to the value added to the research enterprise by patient and subject groups."[16] Importantly, the people who have created these models apparently believe that a notion of the common good can be reclaimed for modern biomedicine, although in the PXE example that conception is limited to those who suffer from a particular genetic condition.

By contributing not just their spit but also the figurative sweat of their brow—the labor they put into completing repeated questionnaires—23andMe customers create corporate "biovalue."[17] I noted in chapter 2 that many analysts of direct-to-consumer genetic testing view the retail genetics companies' biobanks and databases as more valuable than the test-marketing business itself. Information from the spit samples on customers' genotypes is of lesser value, however, without additional phenotype information about the medical conditions from which customers and their relatives have suffered, particularly for genomewide association studies and epigenetic research. That's where 23andWe apparently comes in.

With the contribution of its "active genomes," 23andMe will be able to market their combined genotype-phenotype databases to other firms and to stake lucrative patent claims—for example, if they can link genotypic and phenotypic data to discover new genetic markers. We saw in chapter 2 that this possibility has already come to pass, with the award to the company of a Parkinson's disease genetic patent in June 2012.[18] Biotechnology firms sometimes merely patent defensively against their competitors, but they also use patents as bargaining chips in negotiations and as a means of attracting external finance from banks, venture capital, and other sources.[19] Frequently, patents and the promise they embody are the most crucial and valuable items in a startup firm's portfolio.[20]

"Such patent claims show an agenda that is clearly very different from the image 23andMe is trying to portray to their contributors and to the world at large."[21] In another application (W02011/065982), 23andMe is seeking to patent a method of screening for polymorphisms associated with Parkinson's disease, along with a method of treating the disease using an as yet undiscovered drug.[22] It may seem odd that a company can seek a patent on a drug that hasn't been discovered yet, but that shows just how permissive U.S. patent policy has become.

Although 23andMe is not a drug company, it's apparently in negotiations with pharmaceutical firms to whom its databases are attractive; both Roche and Johnson and Johnson are among the firm's investors. Wojcicki was quoted on the website BusinessInsider.com as saying: "23andMe hopes to partner with more pharmaceutical companies on phase I and phase II trials, which could lead to actual profit. Before then, 23andMe needs to bulk up its DNA database."[23] The firm's commercial strategy rests on a loyal, recontactable customer base and ongoing aggregation of the data that these consumers provide altruistically.[24] (I say altruistically, but there's also a none-too-subtle element of narcissism in surveys titled "Ten Things About You" and "Ten More Things About You," as well as in the injunction to "Find out which traits make you stand out from the crowd." Some commentators think that 23andMe is appealing to an "entrepreneurial self" rather than an altruistic one.)[25]

This model combines the supposedly democratizing power of the Internet with the similar populist and popular appeal of direct-to-consumer genetics, rallying the troops with the now-familiar language of revolutionary fervor. As the firm's blog *The Spittoon* puts it: "Wikipedia, YouTube and MySpace have changed the world by empowering individuals to share information. We believe this same phenomenon can revolutionize healthcare." Interestingly enough, this "revolution" is portrayed as a form of "bowling together"—of getting involved not with your local PTA, Elks, or Junior League but with the Internet community. As the website continues, "We want to provide you the opportunity to connect to and create communities around common interests, affinities and passions."

So far, so Bedford Falls. However, in reality, the value created in part by 23andMe's own laboratories but also in substantial part by its customers' "common interests, affinities and passions" is becoming a private resource for the company and its investors. Like Google, which is one of those backers,[26] 23andMe is "building . . . industrial and commercial empires through the aggregation of data supplied voluntarily and freely by Internet users."[27] And again like Google, 23andMe's core product is the information it has about people who use its services. With an online provider such as Facebook or Google, it's widely suspected that if you're getting the service for free, you're not that company's customer—you're the product. But here the two models part company: with 23andMe's research database, even if you're paying, you're still the product.

Don't get me wrong: communal spirit, altruism, and commitment to scientific research are great things, as are "common interests, affinities and passions." The equally attractive notion of common benefit is invoked by 23andMe in its appeal to participate in scientific progress. The firm consciously creates new sorts of communal identification—the community of other customers as research participants in league with exciting modern biotechnology—to tap into the sense of identification with a wider collective, which underpinned the influential notion of the "gift relationship" developed by the sociologist Richard Titmuss to describe the UK blood donation system.[28] So if 23andMe has created a gift relationship for our times, what's wrong with that?

Although I'm a great fan of the gift relationship when it's genuine, what I find troubling in this instance is the slimness of the likelihood that this communal resource and the patents associated with it will belong to all of us together, or even to the 23andMe customers who contributed their labor and cash to it. I'm not alone in my disquiet: one "active genome" wrote angrily, about the announcement of the genetic patent in June 2012, "I had assumed that 23andMe was against patenting genes and felt in total cahoots all along with you guys. If I'd known you might go that route with my data, I'm not sure I would have answered any surveys."[29]

There probably won't be much "We" about the firm's patent rights or their database, despite the language of sharing. Why do I think that? Legal precedent and commercial pressures argue against it. Here's a prominent example: the Greenberg case, introduced briefly in chapter 2.

Ten years ago, another altruistic group of research participants, who'd also put labor, tissue, dedication, and cash into creating a biobank, found their expectations dashed by a court decision confirming that they had no right to the resource they'd been instrumental in creating. Daniel and Debbie Greenberg had lost their two children Jonathan and Amy to Canavan disease, a rare, genetic, degenerative, and fatal brain disorder primarily affecting families of Ashkenazi Jewish descent. As with other recessive genetic disorders such as cystic fibrosis, parents can carry the harmful allele without realizing it, because they have no symptoms themselves. But if two carriers have children, each child has a one-in-four chance of inheriting two recessive Canavan alleles and of manifesting the condition.

After Jonathan and Amy died, the Greenbergs contacted a genetic researcher, Dr. Reuben Matalon, and asked him to work with them and other Canavan parents. Over a period of thirteen years, they helped him establish a research biobank by contacting over 160 other affected families, arranging for tissue samples from the dead children to be donated to Matalon, soliciting support from the charitable Canavan Foundation, and establishing a database, the Canavan Registry, with valuable additional epidemiological data of the sort gathered by the 23andWe surveys. They and other parents also contributed substantial financial donations to raise funds for Matalon's salary and that of other researchers employed by his hospital in Miami. "All the time we viewed it as a partnership," David Greenberg said afterward.[30]

They were mistaken. Without the Greenbergs' knowledge, the Miami hospital took out a comprehensive patent: U.S. patent number 5,696,635, covering the gene coding for Canavan disease, diagnostic screening methods, and kits for both carrier and antenatal testing. Two years later, the hospital began to collect royalties from the patented genetic test, claiming that it needed to recoup its outlay on research. But although the hospital was indeed a not-for-profit institution, much of that initial outlay had been provided by the research participants, just as in the case of 23andWe. Now other parents would have to pay for screening—which had previously been provided free of charge by the Canavan Foundation, a measure that had helped reduce the incidence of the disease by 90 percent. Furthermore, the hospital used its monopoly patent to impose a number of other conditions, limiting the number of tests performed and the providers licensed to do it.

The Greenbergs lost most of the aspects of the lawsuit that they brought together with other families and the Canavan Foundation, leaving the patent restrictions intact.[31] However, in September 2003 they reached a confidential settlement exempting research doctors and scientists seeking a cure for Canavan disease from having to pay any royalties to the hospital and allowing a number of licensed laboratories to pay no fee for patients having the diagnostic test. In return, they had to agree not to challenge the hospital's ownership of the genetic patent.

Most commentators reckoned that the Greenberg decision favored patent owners such as hospitals or corporations over tissue donors.[32] Yet the argument that patents reward labor and skill doesn't really work for the Greenberg case: Matalon possessed no particular expertise in genetics research. In the opinion of the bioethicist Jon Merz, who worked extensively with the Canavan families, "The only thing that was absolutely required in order to make the discovery was the participation of those families."[33]

Similarly, the only thing that's absolutely required for the 23andMe biobank to possess biovalue that might produce patents and profitable collaborations is the epidemiological data provided at their own cost by the "active genomes." (That, and the original spit samples, of course.) Of course, the firm performs various analyses and provides data to its customers, but we saw in chapter 2 that the reliability of their techniques isn't entirely consistent; nor are the techniques unique to them. By contrast, the "active genomes" are each one of a kind. Particularly for special groups such as Parkinson's patients or African Americans, much of the value must lie in the comparative scarcity value of the contributors: that's presumably why 23andMe has made the business calculation that it's worthwhile to offer such groups free tests.

By winning its first genetic patent, 23andMe is joining the lucrative trend toward private ownership of the ultimate public resource: the human genome.

Even as long ago as 2005, the number of patented human genes was already nearly one in five, with 63 percent of those patents held by private firms.[34] Far from contributing to scientific progress, this privatization of the genome can block scientific and medical practitioners and researchers and worsen patient care,[35] as I'll discuss at greater length in the next section. It also undermines the ultimate We: our human genome, what we all hold in common.

RECLAIMING THE COMMONS

The idea of the genome as the common heritage of humanity underpinned the international agreement reached by scientists in the "Bermuda statement," which declares: "All human genome sequence information from a publicly funded project should be freely available in the public domain."[36] There's also the 1997 UNESCO Universal Declaration on the Human Genome and Human Rights, which states: "In a symbolic sense, the human genome is the common heritage of humanity . . . [and] in its natural state, shall not give rise to financial gain" (articles 1 and 4). So how did we move from this originally communitarian vision for the new genetic biomedicine to the now-dominant personalized medicine paradigm?

Here in the final sections of the book, I want to introduce a new argument: under the pressure of corporate interests and neoliberal policies, we've lost sight of the idea of the commons in biomedicine—but it has only disappeared temporarily from view. Reclaiming biotechnology for the greater good will involve resurrecting the commons. That's a tall order, I know, but moves are already afoot to give us grounds for optimism. The commons has become a focus of activism, from the Occupy Wall Street movement that was in the media spotlight in 2011 to decisions involving private patenting in the U.S. Supreme Court.

Governance of the commons has received substantial attention and analysis in terms of common property in land and the environment.[37] But the no-property-in-the-body rule in law has limited its applicability to the idea of a common property in the genome or human tissue. An exception can be made for James Boyle's wide-ranging application of the commons model to "shamans [traditional knowledge], software [information technology and the open-access movement] and spleens [human tissue]."[38]

Although some attention has focused on the genome, many other aspects of modern biomedicine could, and I think should, be considered a commons.

This is a novel argument, one original to this book. I've been interested in the concept of the commons in biomedicine for some time, primarily in terms of commodification of the body.[39] Yet my thinking has moved on through considering the wide range of examples I've analyzed in this book. I now see additional weight and heft in the concept of the commons, extending beyond the genome, although certainly also relevant there.

Take the notion of herd immunity, encountered in the vaccination cases. When between 85 to 95 percent of the population contribute individually by having themselves immunized, they create a commons: population immunity. They needn't do so out of altruism: as I argued in chapter 6, it's also the correct and prudent decision for each of them as individuals, provided that the side effects don't outweigh the probable benefits. Advocates of the free market ought to accept this argument: it actually mirrors the "invisible hand" model espoused by Adam Smith, in which individual economic decisions produce the best outcome for the collective without any external intervention.

This nexus of thousands or millions of personal decisions then creates a common property in population immunity, which also benefits those incapable of being vaccinated (neonates and the very elderly). The sum is indeed larger than its component parts, and that sum could be considered a commons. Everyone has equal rights to benefit from the intangible commons of population immunity, much as all villagers could claim grazing rights in the traditional land-based commons, but neither commons belongs to any one individual.

Public health affords other examples of a biomedical commons. Life expectancy in the nineteenth century was extended through the creation of such a commons in resistance to smallpox, along with another beneficial commons in clean water and effective sewage, greatly lessening the incidence of cholera and typhus. In modern biomedicine, we can also regard public bodies such as the NIH, medical charities, and governments as having created a form of the commons through the Human Genome Project and other massive public investments in biotechnology.

However, there are two ways in which any commons can be threatened:

1. Individual commoners may endanger the communal resource by taking more than their fair share, or
2. the valuable commons may be enclosed and turned wholly or partially into a private good, depriving the previous commoners of a rightful share in it at all.

First, in the vaccination example, free riders used the argument that population immunity is high enough not to be undermined by their individual decision to withdraw their children from vaccination programs. This reasoning utterly fails the Kantian categorical imperative test, which may be stated (in one version) as acting on a principle that you could will to become a universal law.[40] If everyone followed this free-rider line, population immunity would decline to zero. Since the reasoning relies on population immunity remaining high, it's hopelessly self-contradictory.

Free riding on population immunity exemplifies what Garrett Hardin called "the tragedy of the commons."[41] In Hardin's view, communal property is prone to abuse because everyone who has a share in the common resource is tempted to overuse it. That's why English villages had a pound in which to pen animals whose owners had breached the limits of how many geese or sheep could legitimately be grazed on common land. Commoners could also be tried and fined for overuse of communal rights.

It may seem a bit of a jump from geese and sheep to vaccination, but the parallel is instructive, not just because both involve "herds" but because the tragedy of the commons applies in both cases. Free riders are abusing the resource of herd immunity built up by others: they rely on it to protect themselves and their children, but they don't contribute to it. By declining vaccination, they subtract from the balance, just as surely as greedy commoners who grazed too many sheep or geese lessened the amount of grass available to other people's animals. As animals might decline or even die if insufficient grazing is available, so neonates and the elderly may be particularly vulnerable if population immunity against contagious disease declines.

In the extreme—seen in overgrazed land that succumbs to soil degradation or overused fisheries that never recover their previous plenty, such as the Grand Banks of Newfoundland—the common resource declines so substantially that there is virtually no commons left. That would be the final extreme of the tragedy of the commons. In the vaccination example, population immunity to measles declined so seriously after many parents withdrew their children from MMR vaccination that dangerous epidemics resulted. If that trend were to continue, there would be no commons in population immunity to measles. We're not there yet, by any means, and MMR vaccination levels do now seem to be returning to their pre-1997 levels.[42] But we don't want to go anywhere near there.

Second, there is a link between the way in which free riders take benefits from a commons without having contributed to it and the trend in modern biomedicine toward viewing the human genome or public biobanks as "an open source of free biological materials for commercial use."[43] Over twenty years ago, Justice Broussard warned against that tendency in his minority opinion in the *Moore* case, when he advocated a new commons in human tissue.

> It is certainly arguable that as a matter of policy or morality it would be wiser to prohibit any private individual or entity from profiting from the fortuitous value that adheres in a part of the human body and instead to require all valuable excised body parts to be deposited in a public repository which would make such materials freely available to all scientists for the betterment of society as a whole.[44]

The way in which surplus value is "leveraged" from individual contributions in a commons has been evident in several examples from this book, not just that of 23andMe. Centralized public biobanks or databases for pharmacogenomic research and eventual clinical use can be viewed as a commons from which individual patients should eventually derive benefit. Publicly banked umbilical cord blood can be seen as another example of a commons, a particularly striking one.

Cord blood donated to public banks has far greater clinical value than blood banked privately by the mother for her baby's exclusive use—even to the baby herself. If the child later needs a transplant, she would normally be better off medically by using the blood of others.[45] Additionally, the *ex utero* procedure through which blood is typically collected for public banks is far less risky for mother and baby than the *in utero* method more often used in private cord blood banking. As with the decision to undergo vaccination, there are again both prudential and ethical reasons for the individual to opt for the solution that produces the commons, to elect We over Me. We Medicine isn't just idealistic: in these cases, it wins out on practicality too.

A commons in cord blood is also open to all, regardless of wealth. Whereas ethnic minority parents may be less able to afford private cord blood banks, public banks often make it a priority to offer suitable tissue matches for ethnic minorities through geographically targeted collections.[46] There's little risk of the public resource being overused, as in the tragedy of the commons. By contrast, implementing equality in access to private cord banks poses complicated problems for government intervention and may end up subsidizing private industry at state expense.[47] The way in which public cord blood banks now trade units commercially, however, introduces an element of the market into the pure commons model, revealing one potential weakness of a modern-day biomedical commons. If public banks are tempted to sell too many of their units on the commercial international market, the communal resource could in principle be overused, with the tragedy of the commons becoming an unpleasant prospect.

Although the notion of the genome as the common heritage of humanity and the principle that the human body and its parts shouldn't give rise to commercial gain are ensconced in international agreements such as the "Bermuda statement" and the UNESCO declaration, these instruments afford all of us "genetic commoners" only a weak form of protection. There's no comparable authority to a state agency out there policing the genetic commons. When a not-for-profit agency does act as the authority in charge of a commons—as do public and charitable cord blood banks—it acquires rights to manage that commons. Whether it should have the right to sell individual components of the commons, like individual cord blood units, isn't at all clear, however. Although, like UK Biobank, public cord blood banks have ownership rights in the tissue and data they store, property is regarded in law as a bundle of separate rights.[48] Selling is a separate "stick" in that bundle from administering or managing—just as I have the right to manage my vote as I see fit, casting it as I please, but I don't have the right to sell it.

Precisely because of the tremendous potential value of a commons, there is always a risk of that capital being asset-stripped through sale of parts or the whole. The commons may be turned entirely to a private good, as occurred during the enclosures of common lands in Britain during the eighteenth and nineteenth centuries, entailing the abolition of commoners' traditional rights. Freed of feudal legal restrictions on transfer of ownership and no longer obliged to afford traditional rights to commoners, large landowners could now sell their holdings to raise capital, which helped fund the Industrial Revolution. Meanwhile, the dispossessed commoners had to leave their rural homes and flock to the growing cities, as landless laborers and impoverished factory workers.

The risk then becomes the tragedy of the *anti*commons: when property owners, having enclosed the commons and made it their private possession, use their legal rights to restrict access to a good that would benefit everyone.[49] Scottish lairds who expelled their tenants in favor of sheep or deer, resulting in widespread forced emigration and famine, are one historical example from the enclosure movement. The Canavan case shows that something similar can happen in modern biotechnology, with access to the benefits of diagnostic testing based on a communal resource being denied or limited by a private owner.

The tragedy of the commons is premised on the idea that common property encourages misuse. These two cases, however, show when a commons is privatized in the name of more efficient use, it can also be misused. But we can go further in defending the commons. Unlike land, information is inherently nonrivalrous: more than one person can use it at any time without using it up. The human genome, too, is nonrivalrous. Yet when an artificial scarcity is imposed through restrictive genetic patenting, the tragedy of the anticommons looms.

A NEW GENETIC ENCLOSURE MOVEMENT AND A NEW RAINBOW COALITION

In an influential analysis, the legal scholar James Boyle has argued that biomedicine and information technology are indeed in the throes of a second enclosure movement, in which genetic patenting is the means by which the value of the commons is turned to private benefit.[50] Just as the agricultural enclosures turned common wealth into private possessions, so this strategy attempts to profit from and privatize the public resource of the human genome, both in the Third World and in the wealthier countries.

Far from acting as a democratizing force, however, this process transfers wealth from the public domain to biotechnology venture capital. Facilitated by the absence of individuals' ownership rights in their tissue and by consciously proinvestment patent office policies leading to abuses in wholesale genetic patenting,[51] that transfer has proceeded apace. The genome is Boyle's principal focus, but his analysis could also apply to private cord blood banking as another way in which "things that were formerly thought of as either common property or uncommodifiable are being covered with new or newly extended property rights."[52]

This second threat—that the valuable commons may be enclosed and turned wholly or partially into a private good—is an ever-present temptation, precisely because a commons is such a valuable resource. The 23andMe biobank is that modern anomaly, a corporate commons—created by the labor, cash, and bodily materials of thousands of individuals yet belonging solely to a private firm and its investors or trading partners. But although that combination seems anomalous, huge numbers of the twenty million new entries in biobanks annually belong to private firms.[53]

The argument that the human genome is the common heritage of humanity only prevents genomic information from being appropriated by any nation-state, but in the absence of a binding international treaty, it doesn't rule out appropriation by biotechnology firms.[54] When private biobanks claim the full panoply of the property rights that courts have accorded them,[55] they can exclude even those who labored to create them from ongoing rights and control. What can be done about this new form of genetic enclosure?

One much-respected proposal to restrain this threat and to restore some element of control to those whose contributions created the biobank is the charitable trust model.[56] Although it doesn't give them ownership rights, this model protects biobank contributors by giving them the same status as beneficiaries of a personal trust. And just as trustees are restricted in what they can do with wealth in a trust by the requirement to act in the interest of the beneficiaries, this model would limit the rights of biobank managers to profit as they please from the resource or to sell it on to commercial interests.

The charitable trust model gets around many of the objections against people having any rights in their tissue or genetic data once it's contained in a public or private biobank. Because the contribution of any individual donor to a biobank is minimal, it doesn't seem appropriate for individuals to have full veto powers over how they want their tissue used in research. Additionally, there are problems about recontacting donors (although 23andMe would have fewer problems than, say, UK Biobank, because it has consciously created mechanisms to encourage its "active genomes" to stay active). So the charitable trust model offers joint rather than individual control, effectively creating a new kind of commons.

Nevertheless, the charitable trust model sets out a far more precise roster of duties and entitlements than the rather vague notions of "stewardship" and "custodianship" found in some biobanks' literature. For example, in the version of the model developed by David Winickoff and Larissa Neumann, full disclosure of all pending commercial interests must be made to the tissue donor. A "donor approval committee," drawn from shareholder models in corporate law, could be elected periodically through proxy voting. And if the biobank fails or goes bankrupt—a very real risk in the "easy come, easy go" world of modern biotechnology and venture capital—charitable trust status would prevent its altruistically donated assets from simply passing to the highest bidder or to a creditor.

While critics complain that all this complicated paraphernalia would put donors off donating, they fail to take into account that people are already put off donating when they perceive that their altruism is being exploited by commercialization of their donation.[57] We saw that phenomenon in the angry remarks of the 23andMe customer, who felt so betrayed by the firm's commercial patenting strategy that he was ready to refuse to contribute any more information. If those who donate tissue and time became disenchanted about the possibility of genuine We-ness, they might refuse to donate altogether. The charitable trust model could offer guarantees to keep their altruism alive and reclaim the notion of the common good in biomedicine.

Giving control rights to those whose tissue creates the commons is particularly sensitive and crucial in the case of indigenous peoples. Not only is their biodata frequently of high commercial value; in addition, they typically regard it as a communal holding, not as something any one individual or firm can rightfully claim. Elsewhere, I've discussed such an instance in Tonga,[58] where a local resistance movement overturned an agreement between the health ministry and an Australian firm to collect tissue samples for the purpose of genomic research into the causes of diabetes. Although the disease has a very high incidence in Tonga, and the proposals did include some benefit sharing, Lopeti Senituli, the leader of the resistance movement, urged his people to protect their genetic commons: "Existing intellectual property rights laws favor those with the technologies, the expertise and the capital. All we have is the raw material . . . our blood. We should not sell our children's blood so cheaply."[59] Of course, the Tongans were not literally being asked to sell their children's blood. But there was also a major issue of communal consent: "The Tongan family, the bedrock of Tongan society, would have no say, even though the genetic material donated by individual members would reflect the family's genetic make-up."[60]

Similarly, when the Ojibwe novelist Louise Erdrich was contemplating sending a DNA sample off for genetic analysis, she was reprimanded by family members for failing to consider the impact on them: "It's not yours to give, Louise." We shouldn't underestimate how strong communal notions of We remain—and not only among indigenous peoples. Nor should we underestimate the judicial and political impact that the notion of a genetic commons can wield.

In 2010, the Havasupai tribe of northern Arizona effectively won a legal battle in which they had claimed the right to determine what was done with their genetic data.[61] Like the Tongans, they appealed to traditional communal belief systems in their action against Arizona State University researchers, who had used DNA samples originally gathered in diabetes research for further work on schizophrenia, publishing the results. The tribe protested that they would not have given consent for schizophrenia research, which they viewed as stigmatizing. An award of $700,000 in damages was made to forty-one tribe members by the university, and the tissue samples were returned.

Although this case was settled out of court and thus doesn't leave a binding precedent, it has potential significance not just because of the comparatively low damages, but because of the implication that the gift of tissue doesn't cancel out all the donor's ongoing rights of control. In other decisions, gift of tissue had been held to be final.[62] That's debatable, because in other situations donors do retain ongoing rights in their gifts: for example, a university must use a bequest to build a new library for that purpose alone, not divert it into general operating funds. If the doctrine of gift's finality has really been overturned by the implications of *Havasupai v. Arizona State University*, it will require cataclysmic shifts in biobanks' and biotechnology firms' strategic planning.

Another way of restricting private depredations on the biotechnological commons is to contest restrictive patent rights. Those who favor privatization of the commons through modern biotechnology make exactly the same sorts of arguments as the defenders of enclosure: that it's vital to the march of progress, that it creates efficiencies of production and encourages investment through the release of capital, even that it saves lives (in the enclosure instance, by raising the total amount of foodstuffs produced, although that says nothing about their distribution). This is the classical argument behind patenting genes: that research won't proceed unless researchers and their employers have secure protection against their inventions being stolen by rivals.

Against this argument, research and clinical treatment can be blocked by restrictive monopoly patents, as in the Herceptin example from chapter 3. The effect of patents is to take genes out of the genomic commons, although the impact of that can be softened with compulsory licensing to other firms or similar measures to restrict the patent holder's absolute right to sell products derived from that gene at a monopoly price.

So while there's cause for optimism in some court cases that acknowledge indigenous people's belief and common rights in their genetic information and tissue, the really valuable holdings—in lucrative patents—may not be so easy to reclaim from corporations and researchers. The Myriad Genetics case, discussed at greater length in chapter 2, is a case in point. But whatever the outcome that case has created a kind of communal action. The plaintiffs included the American College of Medical Genetics, the American Society for Clinical Pathology, the College of American Pathologists, individual physicians, the Boston Women's Health Book Collective, and several individual women who claimed that their clinical condition had worsened because of their insurers' unwillingness to cover the $3,000+ cost of the BRCA1/2 test. *Amicus curiae* (friend of the court) advisory briefs were submitted on behalf of civic groups ranging from the Southern Baptist Convention to the Canavan Foundation. Such a rainbow coalition could itself be seen as a kind of commons, with a doubly We Medicine character: it unites some very unusual bedfellows in pursuit of a goal that can be seen as restoring the genetic commons.

In this final chapter, we've encountered a number of collective movements attempting to create a new spirit of togetherness in modern biomedicine: the Havasupai, PXE International, the coalition of plaintiffs in the *Myriad* case, Lopeti Senituli and his fellow Tongans, advocates of the charitable trust model, and "genetic commoners" of many sorts. The notion of We is alive and well in modern biomedicine, although it faces substantial political and economic hurdles. Me Medicine isn't the only possibility. Nor is it an inevitability. If we do choose to embrace personalized medicine, it should only be after a thorough review of the evidence and a careful analysis of the social landscape in which we're making that choice, since the scientific evidence alone *doesn't* dictate that you have to be ready to embrace the personalized paradigm.

In offering that reality check in this book, I've given you the chance to decide between Me and We Medicine or perhaps to thread your own way between the two. That may sound a bit flat compared to the high-pitched trumpet reveilles of most books on personalized medicine, but you're not in the army. I've been a bit tough on choice throughout this book, but here's where you really can choose. Although I'm glad of genuinely beneficial Me Medicine developments where there really is evidence for them, I'm sure you've guessed my choice: it's We for me whenever I can.

NOTES

1. A REALITY CHECK FOR PERSONALIZED MEDICINE

1. Collins 2010:xxiv–xv.
2. Fidanboylu 2011.
3. Arribas-Ayllon 2010; Genewatch UK 2010.
4. Wade 2010.
5. Kimmelman 2010.
6. United Health 2012:3.
7. United Health 2012:4.
8. Hogarth and Martin 2012.
9. RCOG 2002, 2006.
10. E.g., Fleming 2008.
11. Sandel 2004.
12. Reported in Mnookin 2011:14.
13. Sarojini et al. 2011; Sengupta et al. 2011.
14. Jegede 2007.
15. Hwang 2004, 2005.
16. Paik 2007.
17. Dickenson 2006; Baylis 2009.

18.   E.g., Brown 2006.

19.   Quoted in White 2011.

20.   Quoted in White 2011.

21.   Ginsburg and Huntington 2009.

22.   Nuffield Council on Bioethics 2010.

23.   Genewatch UK 2010.

24.   Gilbert 2010.

25.   Personalized Medicine Coalition 2009.

26.   Topol 2012:viii.

27.   Personalized Medicine Coalition 2009:3.

28.   Collins 2010:xxiv–xv.

29.   Quoted in Duncan 2009:34.

30.   Hebert et al. 2008; Nuffield Council 2010.

31.   Robson et al. 2010.

32.   Douglas 1966.

33.   Waldby and Mitchell 2006:54.

34.   Waldby and Mitchell 2006:55.

35.   Hermitte 1996.

36.   Royal College of Obstetricians and Gynaecologists 2001, 2006.

37.   Wallace 2009.

38.   Sparrow 2011.

39.   Campbell 2012.

40.   Fleck 2010.

41.   President's Council of Advisors on Science and Technology 2008:s2.

42.   Shanks 2011.

43.   Nordgren 2010.

44.   Nordgren 2010.

45.   Twenge and Campbell 2009:67.

46.   Duncan 2009:6.

47.   Twenge and Campbell 2009:240.

48.   Twenge and Campbell 2009:67–68.

49.   Nelkin and Lindee 1995:41–42.

50.   Fowler and Dawes 2007.

51.   Knafo et al. 2008.

52.   Lippman 1998.

53.   Putnam 2000:27.

54.   Putnam 2000:17.

55.   Quoted in Brockes 2012:31.

56. Putnam 2000:18–19.

57. Quoted in Putnam 2000:22.

58. Quoted in Hedges 2010:156.

59. Frank 2012a.

60. Porter 2007.

61. Brody 2009:139.

62. Sulston and Ferry 2003:309–310.

63. Klein 2008:15.

64. Hall 2011.

65. Sandel 2012.

66. Hall 2011:12.

67. Krimsky 2011.

68. Office of Science and Technology Policy 2012.

69. White House 2012:6.

70. Sigurdsson 2001.

71. Executive Order of April 21, 2010: section 2ii.

72. Nuffield Council on Bioethics 2010: section 2.29.

73. Steinbrook 2005.

74. Angell 2005.

75. http://www.navigenics.com.

76. E.g., Darnovsky 2008.

77. Hastings Center Report 2011.

78. Dickenson 2008.

79. E.g., *Moore* 1990; *Greenberg* 2003.

80. Beauchamp and Childress 2009.

81. Beauchamp and Childress 2009; Gillon 1986.

82. Hedgecoe 2004; Fox and Swazey 2008.

83. Dickenson, Huxtable, and Parker 2010:191.

84. Widdows 2011:13.

85. Sherwin 1992; Tong et al. 2004.

86. O'Neill 2002.

87. Callahan 1984, 2003; Parker 1999, 2012; Etzioni 2011.

88. Jonas 1982.

89. E.g., Frankfurt 1971; Dworkin 1988.

90. Dickenson 2011.

91. Tupasela 2010.

92. Darnovsky 2010.

93. Johnson 2010.

94. Savulescu 2003:138–139.
95. Widdows 2011:15–16.
96. Battin et al. 2008.
97. Meghani 2010.
98. Darnovsky 2010.
99. Rawls 1971.
100. Pateman and Mills 2007:2.
101. Dickenson 1997.
102. Pateman and Mills 2007:14.

2. "YOUR GENETIC INFORMATION SHOULD BE CONTROLLED BY YOU": PERSONALIZED GENETIC TESTING

1.   Nordgren 2010.
2.   18 Vermont State Acts 9330; Massachusetts bill, section 1.
3.   Quoted in Genomics Law Report 2011.
4.   UNESCO 1997.
5.   Gruber 2011.
6.   Quoted in McGowan et al. 2010:264.
7.   Malak and Daar 2012.
8.   Sulston and Ferry 2003:212.
9.   Quoted in Gruber 2011:1.
10. Collins and McKusick 2001.
11. Brooks and Tarini 2011.
12. Roberts 2011:204.
13. Borry, Cornel, and Howard 2010.
14. Borry, Cornel, and Howard 2010.
15. Wade 2010.
16. Paynter et al. 2010.
17. Arribas-Ayllon 2010.
18. Carey 2011.
19. Carey 2011:42.
20. Venter et al. 2001.
21. Paynter et al. 2010.
22. Arribas-Ayllon 2010.
23. McGuire et al. 2009.
24. President's Council of Advisers on Science and Technology 2008:27.
25. Hall et al. 2002.

26. Quoted in Wallace 2009:81.

27. Roe 1988.

28. Genewatch UK 2010:8.

29. Wallace 2009.

30. Genewatch 2010:2.

31. Gundle et al. 2010.

32. Gundle et al. 2010:974.

33. Dondorp and de Wert 2010.

34. Roberts 2011:208.

35. Angrist 2010.

36. Duncan 2009; Richards 2010.

37. Kalf et al. 2011.

38. Goldstein and Cavalleri 2005.

39. McGowan et al. 2010:263.

40. McGowan et al. 2010:263–264.

41. Eng et al. 2010.

42. Richards 2010:306.

43. Dondorp and de Wert 2010:10.

44. Quoted in Singer 2011.

45. Robson et al. 2008.

46. Robson et al. 2010:894.

47. Salzberg and Pertea 2010.

48. Wilkie 2009.

49. E.g., Foster et al. 2009.

50. Robson et al. 2010:896, 894.

51. Genomics Law Report (March 24, 2011).

52. Bloss et al. 2012.

53. McGowan et al. 2010.

54. Saunders 2012.

55. Institute of Medicine 2012.

56. GAO 2010.

57. Quoted in Saunders 2012.

58. Hamburg and Collins 2010.

59. Collins 2010:xxiv–xxvv.

60. Knoppers et al. 2010.

61. Roberts 2011:204.

62. Lancet 2010.

63. Nuffield Council on Bioethics 2010:226, 142.

64. Cooper 2007.

65. Decker and Bruun 2012.

66. Skloot 2010.

67. Sulston and Ferry 2003:309.

68. Arribas-Ayllon 2010; Borry 2010; Roberts 2011.

69. Quoted in Roberts 2011:209.

70. Dondorp and de Wert 2010.

71. Waldby and Cooper 2010; Brown et al. 2010.

72. *Greenberg et al. v. Miami Children's Hospital Research Institute* 2002.

73. *Moore v. Regents of University of California* 1990:506 (emphasis in original).

74. Both quoted in Andrews 2006:401.

75. Cited in Dickenson 2008:130–131.

76. Quoted in Andrews 2006:398–399.

77. Mason and Laurie 2011:460.

78. *Yearworth v. North Bristol NHS Trust* 2009: para. 45 (emphasis added).

79. Association of Molecular Pathology 2010.

80. 18 VSA §9336(a).

81. Genomics Law Report 2011.

82. 18 VSA §9336(c).

83. Quoted in Roberts 2011:205.

84. Winickoff 2003; Boggio 2005; Hoppe 2010.

85. Wallace 2009.

86. Quoted in Dickenson 2010.

87. Brooks and Tarini 2011.

88. Hebert et al. 2008:187.

89. Quoted in Kahn 2012.

90. Kahn 2011.

91. Paynter et al. 2010.

92. Richard 2010:306.

93. Alther 2007.

94. Roberts 2011:230.

95. Mykitiuk 2000.

96. Roberts 2010:13.

97. Klein 2007; Mirowski 2011; Krimsky 2011.

98. Mazzucato 2012:26.

99. Hogarth 2009.

100. E.g., Frank 2011.

101. E.g., Annas 2010; Fox and Swazey 2008; Etzioni 2011.

3. PHARMACOGENETICS: ONE PATIENT, ONE DRUG?

1. Quoted in Mukherjee 2011:37.
2. Mukherjee 2011:37.
3. Rieff 2008; Mukherjee 2011:306.
4. Mukherjee 2011:362, 231.
5. In this chapter, I generally use the term "pharmacogenetics" rather than "pharmacogenomics." Most commentators view them as interchangeable (e.g., Goldstein et al. 2003), but "pharmacogenetics" is probably more familiar, as it is the older term and has been in use since its coinage by Friedrich Vogel in 1959 (Nebert et al. 2008). Other authors also prefer the term pharmacogenetics, meaning clinical testing of genetic variation, rather than pharmacogenomics, focusing "more on broader application of genomic technologies to new drug discovery" (Lee and McCleod 2011:15).
6. Goldstein et al. 2003.
7. Thornhill 2008.
8. Wadman 2011.
9. Curtis et al. 2012.
10. Mukherjee 2011:38.
11. E.g., Goldstein et al. 2003.
12. PHG Foundation 2010.
13. Sosman et al. 2012.
14. Mukherjee 2011:197.
15. Gerlinger et al. 2012:883.
16. Link et al. 2011.
17. Worthey et al. 2010.
18. Thornhill 2008; Laberge and Burke 2008.
19. Samuel 2010.
20. Lorizio et al. 2011.
21. Mukherjee 2011:418, 273.
22. Quoted in Fathimathas 2010.
23. Callaway 2012.
24. Morley 2011.
25. Linn 2012.
26. Sequist et al. 2011.
27. Maher 2011.
28. Heger 2010.
29. Lyons 2012.

30. Worthey et al. 2011:255.

31. Bainbridge et al. 2011.

32. Gura 2012.

33. Maxmen 2011.

34. Roberts 2011; Kahn 2007; Obasogie 2010.

35. Kramer et al. 2009.

36. Calderon-Margalit et al. 2009.

37. Jacobs et al. 2001.

38. Thornhill 2008.

39. Neary 2010.

40. Quoted in Wijlaars 2012.

41. Arranz and Leon 2007:707.

42. Goldacre 2008; Mukharjee 2011.

43. Quoted in Harper 2011:2.

44. Ginsburg and Willard 2009.

45. Wadman 2011.

46. Lee and McLeod 2011:15.

47. Quoted in Hedgecoe 2007:1.

48. Lee and McLeod 2011:15, 19.

49. Robson et al. 2008.

50. E.g., Cudkowicz 2010.

51. Pollack 2010.

52. Nebert et al. 2008:187.

53. Quoted in Harper 2011:3.

54. Dion-Labrie et al. 2010.

55. Dion-Labrie et al. 2010.

56. Braithwaite 1963; Kahneman et al. 1982.

57. Dickenson 2003:107.

58. Calabresi and Bobbitt 1978:133.

59. Sanders and Dukeminier 1968.

60. Harris 1985.

61. Dickenson 2003:107–108.

62. E.g., Harris 1985.

63. Goodwin 2006:44.

64. Dion-Labrie et al. 2010.

65. Lee and McLeod 2011:22.

66. Hatzis et al. 2011; Stone 2011.

67. Ginsburg and Willard 2009.

68. Quoted in Stone 2011.

69. You 2011.

70. Sjogren 2010.

71. Sharp 2012.

72. Quoted in Roberts 2011:151.

73. Ling and Raven 2005:s8.

74. Dickenson 2012.

75. Ling and Raven 2005.

76. Quoted in Roberts 2011:179.

77. Roberts 2011:166, 168.

78. I use the term "ethnic minorities" for want of a more readily recognized alternative, although I realize that in areas such as New Mexico, Anglos have long been in the minority, as they will eventually be in the country as a whole. "Persons of color" is the alternative on which I sometimes draw.

79. Quoted in Roberts 2011:164.

80. Kahn 2007.

81. Roberts 2011:185.

82. Cited in Roberts 2011:180.

83. Based on Roberts 2011; Kahn 2004, 2007.

84. Carson et al. 1999.

85. Brody 2009:150–151.

86. Brooks and King 2008.

87. Quoted in Roberts 2011:171.

88. Kahn 2004:11.

89. Quoted in Roberts 2011:174.

90. Roberts 2011:172.

91. Quoted in Roberts 2011:183.

92. Kahn 2004.

93. Washington 2011.

94. Roberts 2011:198.

95. Tate and Goldstein 2004.

96. Brooks and King 2008.

97. Brooks and King 2008.

98. Nature Biotechnology 2005:903.

99. Kahn 2004:11.

100. Condit et al. 2003; Bates et al. 2004.

101. Wallace 2010.

102. Mukherjee 2011.

103. Andrews 2002; Secretary's Advisory Committee on Genetics, Health, and Society 2010.

104. Jensen and Murray 2005.

105. Cooper 2007.

106. Chiang and Milton 2011:895.

107. Clarke and Everest 2006.

108. Beach et al. 2005.

109. Goldstein et al. 2003.

110. Chiang and Milion 2011:895.

111. Dutfield 2009:344.

112. Chiang and Milion 2011.

113. Kwak et al. 2010.

114. Chiang and Milion 2011:895.

115. Chiang and Milion 2011:895–896.

116. Chiang and Milion 2011:895.

117. Dion-Labrie et al. 2010.

118. Fauser et al. 2010.

119. Nebert et al. 2008.

120. Geddes 2011.

4. "YOUR BIRTH DAY GIFT": BANKING CORD BLOOD

1. Gluckman et al. 1989.

2. McKenna and Sheth 2011.

3. American Academy of Pediatrics 2010.

4. Busby 2010.

5. Cranage 2011.

6. Yarwood 2012.

7. Ballen 2010.

8. McKenna and Sheth 2011:264.

9. Ballen 2010:11; Kaimal 2009.

10. Gluckman 2001.

11. American Academy of Pediatrics 2010.

12. American Academy of Pediatrics 2007; Ballen 2010; RCOG 2001, 2006; Greek National Bioethics Commission 2007; Flegel 2009; Thornley 2009; Bienvault 2010.

13. Hardin 1968.

14. Brown 2011.

15. Gènéthique 2011a.

16. Le Brocq 2008.

17. Pacificord 2011.

18. Examples from RCOG 2006.

19. Fox et al. 2007.

20. Waldby and Mitchell 2006:125.

21. Ballen 2010:8.

22. Wiemels et al. 1999.

23. Rocha 2000.

24. Ballen 2010:12.

25. Barker et al. 2005.

26. Ballen 2010:8.

27. Ballen 2010:11.

28. Ferreria et al. 1999.

29. Devine 2010.

30. Copelan et al. 2009.

31. Haller et al. 2008.

32. Ballen 2010:12.

33. Locatelli et al. 2003.

34. McKenna and Sheth 2011.

35. Prasad et al. 2008.

36. Chatterton 2010.

37. Ma et al. 2005.

38. Cohn 2008.

39. Ballen 2010.

40. Zhao et al. 2012.

41. Warwick and Armitage 2004:995.

42. Ecker and Greene 2005:1281.

43. Ediezen 2006.

44. Kaimal et al. 2009.

45. Thornley et al. 2009.

46. Dickenson 2003:89.

47. Flegel 2009.

48. Downey and Bewley 2009:1; the excerpted passage quotes Rogers et al. 1998.

49. World Health Organization 2007:15.

50. RCOG 2006.

51. Fisk and Atun 2008.

52. Bewley 2011.

53. Hutchon 2008.

54. Rabe et al. 2004; Hutton and Hassan 2007; Weeks 2007; Downey and Bewley 2009; Hutchon 2010.

55. RCOG 2009:1.

56. Tolosa et al. 2010; Paxman 2010.

57. Downey and Bewley 2009:2.

58. Mercer 2001.

59. Van Rheenen and Brabin 2004.

60. Chaparro et al. 2006.

61. Hutchon 2008, original emphasis.

62. RCOG 2009.

63. RCOG 2009:4.

64. Downey and Bewley 2009.

65. Cotter et al. 2001.

66. Walsh 1968.

67. Darwin 1801:302.

68. RCOG 2001, 2006.

69. Davey et al. 2004.

70. Tromp 2001.

71. Laskey et al. 2002; Warwick and Armitage 2004.

72. Brown 2011.

73. Brown et al. 2011.

74. Beatty, Mori, and Milford 1995.

75. Goodwin 2006.

76. Kollman 2004.

77. Davey et al. 2004.

78. Brown et al. 2011:15.

79. Brown et al. 2011:2.

80. Brown 2011.

81. Busby 2010:25.

82. *Sidaway v. Board of Governors of the Bethlem Royal Hospital* 1984.

83. Busby 2010:24–25.

84. Moise 2005; *Christopher v. Pharmastem Therapeutics Inc.* 2008.

85. Waldby and Cooper 2010:3.

86. Dickenson 1997; Dodds 2003; McLeod and Baylis 2006; Ikemoto 2009.

87. E.g., Waldby and Mitchell 2006.

88. E.g., Munzer 1999; Annas 1999; Haley, Harvath, and Sugarman 1998; Kirschenbaum 1997; Vawter 1998.

89. Meyer, Hanna, and Gebbie 2004.

90. RCOG 2001, 2006.

91. Leigh 2005.

92. RCOG 2006: section 6.

93. Munzer 1999; Munzer and Smith 2001:205.

94. E.g., *Moore v. Regents of University of California* 1990; *R. v. Kelly* 3 All ER 741; *Greenberg et al. v. Miami Children's Hospital Research Institute et al.* 2003.

95. *Washington University v. Catalona* 2008.

96. Brown 2011.

97. Virgin Health Bank 2011.

98. PacifiCord 2011.

99. Kaimal 2009.

100. CCNE 2002, 2012.

101. CCNE 2012:19, translation mine.

102. Gènéthique 2011b, 2011c.

103. Carvel 2005.

104. Cooper 2007.

105. Dickenson 2006, 2008, 2013.

5. ENHANCEMENT TECHNOLOGIES: FEELING MORE LIKE MYSELF

1. E.g., Savulescu 2001.

2. Habermas 2003:63.

3. Buchanan 2011:23.

4. STOA 2009:8; emphasis added.

5. Rose 2007:17.

6. Partridge et al. 2011.

7. Buchanan 2011:11–12.

8. Collins 2010:xv.

9. Benet 1927:336.

10. Buchanan 2011:8.

11. E.g., Bostrom and Sandberg 2009.

12. A subscriber to BeautyAddiction.com; quoted in Kuczynski 2006:82.

13. Triggle 2012.

14. Kimmelman 2010.

15. Newman 2010; Roussel and Lechner 2011.

16. Royal Society 2012.

17. Chatterjee 2004.

18. McCabe et al. 2005; Gould 2009; Cakic 2009.

19. Larriviere et al. 2009.

20. Warren et al. 2009.

21. Larriviere et al. 2009:1407.

22. Martin et al. 2012.

23. Grimley Evans et al. 2006.

24. Malouf and Grimley Evans 2008.

25. Lethaby et al. 2008.

26. Lim et al. 2006.

27. Angevaren et al. 2008.

28. Gilbert 2009.

29. Boot et al. 2009.

30. Yesavage et al. 2001.

31. Hall 2003.

32. Dickenson and Vineis 2002; Ter Meulen and Dickenson 2002.

33. Caldwell et al. 2000.

34. Kimberg et al. 1997; Farah et al. 2004.

35. Tang et al. 1999; Tang et al. 2001.

36. Chatterjee 2004:968.

37. Chatterjee 2004:968.

38. Pitman et al. 2002.

39. Farah et al. 2004.

40. Lynch 2002; Tully 2003.

41. Glannon 2007.

42. Glannon 2007:2.

43. LeDoeuff 2000:80.

44. Kimmelman 2010.

45. Sasaki et al. 2009.

46. Rasko and Jolly 2006:19.

47. Schatten and Mitalipov 2009.

48. Sasaki 2009:524, table 1.

49. Schatten and Mitalipov 2009:516.

50. Baylis 2009.

51. Wright and Burton 2008.

52. E.g., Chan and Harris 2008.

53. Dickenson 2006.

54. Savulescu 2002.

55. Savulescu 2001:415, 421, 425.
56. Hathaway et al. 2007.
57. Orkin and Motulsky 1995:1.
58. Kimmelman 2010:1.
59. Baylis and Robert 2006.
60. Kimmelman 2010:10.
61. Engelhardt 1996; Walters 1997; Evans 2002; Chapman and Frankel 2003; Rasko et al. 2006.
62. Davies 2009:228.
63. Schermer 2008.
64. Bess 2011.
65. Chan and Harris 2008; Buchanan 2011:38.
66. Parens 2005.
67. From Kuczynski 2006:67.
68. Elliott 2004:3.
69. Dickenson 2008:136.
70. Elliott 1998, 2004.
71. See, e.g., Sartre 2003.
72. President's Council on Bioethics 2002.
73. Quoted in Sample 2012.
74. Buchanan 2011:75.
75. Buchanan 2011:95 (emphasis in the original).
76. Spring 1971; Illich 1995.
77. Harris 2007; Brassington 2010.
78. Sandel 2007:26–27.
79. Annas 2010:252.
80. Buchanan 2011:221.
81. Annas 2010:257.
82. Fukuyama 2002:172.
83. WHO 2002.
84. Daniels 2008:45.
85. George 2012:44.
86. Daniels 2008:310.
87. Elliott 2004.
88. Gould 2009.
89. Greeley et al. 2008.
90. Moreno 2006; National Research Council 2008, 2009; Burnam-Fink 2011.
91. Royal Society 2012.

92. Douglas 2008.
93. Persson and Savulescu 2008.
94. Fenton 2010; Harris 2010.
95. Huxley 1994:217.
96. E.g., Crockett 2009; Rogers 2012.
97. Crockett 2009:1176.
98. E.g., Harrosh 2012.
99. Royal Society 2012: box 4.
100. Agar 2004.
101. E.g., Harris 2007:3.
102. Skloot 2010.
103. Sandel 2004.
104. Rawls 1971.
105. Sandel 2004.
106. Buchanan 2011; McConnell 2010.
107. Wasserman 2003.
108. Mehlman and Botkin 1998.
109. Gould 2009.
110. Vance 2010.
111. Sulston and Ferry 2003:309–310.
112. Dutfield 2009:167.
113. Dutfield 2009:175, Johnson 2012.
114. Elliott 2004; Lamkin 2011.
115. Farah et al. 2004:422.
116. STOA 2009:108.
117. Vance 2010.
118. Partridge et al. 2011.
119. Parens 2005.
120. Gould 2009.

6. "THE ANCIENT, USELESS, DANGEROUS, AND FILTHY RITE OF VACCINATION": PUBLIC HEALTH, PUBLIC ENEMY?

1. Caplan 2011.
2. Allen 2007.
3. Allen 2007:15.
4. Mitchell 1907; quoted in Allen 2007:99.

5. Prainsack 2011.

6. Poland and Jacobson 2011:97.

7. Colgrove 2010:12.

8. I am grateful to Richard Moxon, emeritus professor of pediatrics and former head of the molecular infectious diseases group at the University of Oxford, for this example.

9. Jegede 2007.

10. Quoted in Jegede 2007:e75.

11. Margaret Chan; quoted in Annas 2010:230.

12. Blume 2006.

13. Allen 2007:71ff.

14. Allen 2007:97.

15. Allen 2007:102.

16. Allen 2007:103.

17. Schwartz 2012.

18. Mnookin 2011:8. Emphasis mine.

19. Allen 2007:17.

20. Lees et al. 2005.

21. Mnookin 2011:307.

22. Sears; quoted in Mnookin 2011:279.

23. Battin et al. 2008.

24. Turner et al. 2010.

25. Pywell 2000.

26. Battin et al. 2008:308.

27. MacFarlane 2009.

28. Landry's paralysis or Guillain–Barré-Strohl syndrome is an acute polyneuropathy, a disorder affecting the peripheral nervous system. The most typical symptom is weakness and paralysis beginning in the feet and hands and migrating toward the trunk; some subtypes cause change in sensation or pain and can cause life-threatening complications, in particular if the breathing muscles are affected.

29. Tremblay et al. 2010.

30. World Health Organization 2010.

31. Annas 2010.

32. Cook 2010.

33. Letter of June 3, 1850; in Hudspeth 1993: no. 885.

34. Mnookin 2011:26.

35. RCOG 2001.

36. Annas 2010:209.
37. Annas 2010:17.
38. Battin et al. 2008:315.
39. Poland and Jacobson 2011:98.
40. PCAST 2009:v.
41. PCAST 2009:vi.
42. Murray et al. 2007.
43. Doshi 2007.
44. Doshi 2008.
45. Berry 2011.
46. Nicoll 2010.
47. Jefferson et al. 2009.
48. Kitching et al. 2010.
49. Johnson 2010.
50. Nicoll 2010.
51. Towers and Feng 2009.
52. Johnson 2010.
53. Caplan 2011.
54. Cohen and Carter 2010.
55. Nicoll 2010:2.
56. Rubin 2010.
57. Turner et al. 2010.
58. Goldacre 2008.
59. White 2010.
60. Rubin et al. 2009.
61. Rubin 2010.
62. Nicoll 2010; Berry 2011.
63. Battin et al. 2008:331.
64. Read 2010.
65. Battin et al. 2008:333.
66. Ungchusak et al. 2005.
67. Battin et al. 2008:333–334.
68. Fouchier et al. 2012; Malakoff and Enserink 2012.
69. Cohen 2012.
70. Harris, Mauer, and Kellerman 2010.
71. Fidler and Gostin 2011.
72. Jensen and Murray 2005; Dickenson 2008.
73. Fidler and Gostin 2011:200.

74. Colgrove 2010:11.

75. Lees et al. 2005.

76. Delden et al. 2008.

77. Towers and Feng 2009.

78. Harris and Holm 1995.

79. Battin et al. 2008.

80. Quoted in Offit 2007:96.

81. UK Health Protection Agency 2011.

82. Roxby 2011.

83. Offit 2007:xix.

84. Colgrove 2006a.

85. Jacobson et al. 2007.

86. Wakefield et al. 1998. The full list of authors was: A. J. Wakefield, S. H. Murch, A. Anthony, J. Linnell, D. M. Casson, M. Malik, M. Berelowitz, A. P. Dhillon, M. A. Thomson, P. Harvey, A. Valentine, S. E. Davies, and J. A. Walker-Smith.

87. Wakefield et al. 1998:640.

88. Chen and DeStefano 1998.

89. Wakefield 1998.

90. Deer 2004.

91. Wakefield 2012.

92. Murch et al. 2004.

93. Allen 2007:393.

94. Allen 2007:393.

95. Offit 2007:xxi.

96. Offit 2007:xxvii.

97. Mnookin 2011:180.

98. E.g., Stehr-Green et al. 2003; Madsen et al. 2003; Hviid et al. 2003; Heron and Golding 2004.

99. Mnookin 2011:122.

100. Huffington 2002.

101. Price et al. 2010; Thompson et al. 2007; Fombonne 2006.

102. Schechter and Grether 2008.

103. Offit 2007:174.

104. Quoted in Boseley 2010:4.

105. Wakefield 2012:1.

106. Wakefield 2012:2.

107. Doja and Roberts 2006.

108. American Academy of Pediatrics 2010.

109. Smith and Woods 2010.

110. Budzyn et al. 2010.

111. Uchiyama et al. 2007; DeStefano et al. 2004; Madsen et al. 2002.

112. Smeeth et al. 2004.

113. Hornig et al. 2008.

114. Black et al. 2002.

115. Peltola 1998.

116. Lingam et al. 2003.

117. Lingam et al. 2003:669.

118. Epstein and Huff 2010:215.

119. Haug 2008.

120. Brotherton et al. 2011.

121. Tomljenovic and Shaw 2012:1.

122. Sawaya and Smith-McCune 2007.

123. Braun and Phoun 2010:52.

124. Epstein and Huff 2010:214 (emphasis in the original).

125. Braun and Phoun 2010.

126. John and Shah 2011.

127. Ramogola-Masire 2010:96.

128. Ramogola-Masire 2010:98.

129. Smith 2008.

130. Livingston et al. 2010:235.

131. Epstein 2010.

132. Hoffman 2008.

133. Reich 2010:171.

134. Aronowitz 2010.

135. Colgrove 2006b.

136. Battin et al. 2008:480.

137. Nuffield Council on Bioethics 2007:57.

138. Colgrove 2006b; Epstein and Huff 2010:220.

139. Quoted in Reich 2010:173.

140. Livingston et al. 2010:238.

141. Sarojini et al. 2010:28.

142. Sarojini et al. 2011:18.

143. Srinivasan 2011:73.

144. Sarojini et al. 2010.

145. Larson et al. 2010:572.

146. Srinivasan 2011:74.

147. Sarojini et al. 2010:30.

148. Garland et al. 2011:104.

149. Ramogola-Masire 2010:93.

150. Jayakrishnan 2011.

151. Larson et al. 2010.

152. Sengupta et al. 2011.

153. Jayakrishnan 2011:107.

154. Indian Parliament 2012:9, section 2.2.

155. Jayakrishnan 2011:107.

156. Jayakrishnan 2011:108.

157. Jayakrishnan 2011:108.

158. Douglas 1966.

159. Allen 2007:57.

160. Murphy 2004.

161. Offit 2007:215.

162. Mnookin 2011:273.

163. Allen 2007:428.

164. Allen 2007:425.

165. Prince 1977.

166. Allen 2007:426.

167. Jayakrishnan 2011:107.

168. Allen 2007:433.

169. Quoted in Kahn 2011.

170. Levin 2012.

171. Quoted in Allen 2007:438.

172. Quoted in Mnookin 2011:200.

7. RECLAIMING BIOTECHNOLOGY FOR THE COMMON GOOD

1. Levin 2012.

2. Frank 2012b: l. 317.

3. Frank 2012b:28.

4. Frank 2012b:30.

5. Frank 2012a: l. 1095.

6. Sandel 2012.

7. Frank 2012: l. 1166.

8. Collins 2010:xxiv.

9. Quoted in Harris et al. 2012.

10. HUGO 2000.

11. Hayden 2007.

12. Winickoff 2007.

13. Terry 2002.

14. Winickoff 2007:450.

15. Winickoff 2007.

16. Merz et al. 2002:969.

17. Mitchell and Waldby 2010; Palsson 2009.

18. Saunders 2012.

19. Salter 2007.

20. Cooper 2007.

21. Sterckx and Cockbain 2012:25.

22. Sterckx and Cockbain 2012:21.

23. Quoted in Sterckx and Cockbain 2012:26.

24. Harris et al. 2012.

25. Tutton and Prainsack 2011.

26. Sterckx and Cockbain 2012:26.

27. Proulx et al. 2011:10.

28. Titmuss 1971; Hayden 2007.

29. Anonymous comment on "The Spitoon" blog (June 12); quoted in Saunders 2012.

30. Quoted in Gorner 2000:3.

31. *Greenberg v. Miami Children's Hospital* 2003.

32. E.g., Anderlik and Rothstein 2003.

33. Quoted in Gorner 2000:3.

34. Jensen and Murray 2005.

35. Andrews 2002; Cook-Deegan and Heaney 2010.

36. HUGO 1996.

37. Daly and Cobb 1989; Ostrom 1990; Harris 1996; Shiva 1997.

38. Boyle 1996.

39. Dickenson 2003b, 2005, 2007, 2008, 2009.

40. Kant 1949.

41. Hardin 1968.

42. UK Health Protection Agency 2011.

43. Waldby and Mitchell 2006:24.

44. *Moore v. Regents of University of California* 1990:172.

45. RCOG 2001, 2006.

46. US National Academies 2005; Warwick and Armitage 2004.

47. Zilberstein et al. 1997.
48. Honoré 1987.
49. Heller 1998.
50. Boyle 1996, 2003.
51. Ayme et al. 2008.
52. Boyle 2003:37.
53. Andrews 2005.
54. Knoppers 2005.
55. E.g., *Washington University v. Catalona* 2007.
56. Gottlieb 1998; Winickoff and Winickoff 2003; Otten et al. 2004; Boggio 2005; Winickoff and Neumann 2005.
57. Levitt and Weldon 2005.
58. Dickenson 2007, 2008.
59. Senituli 2004:3.
60. Senituli 2004:3.
61. Harmon 2010.
62. E.g., *Washington University v. Catalona* 2006.

REFERENCES

Agar, Nicholas. 2004. *Liberal Eugenics: In Defence of Human Enhancement*. Oxford: Blackwell.

Allen, Arthur. 2007. *Vaccine: The Controversial Story of Medicine's Greatest Lifesaver*. New York: Norton.

Alther, Lisa. 2007. *Kinfolks: Falling Off the Family Tree. The Search for My Melungeon Ancestors*. New York: Arcade.

American Academy of Pediatrics. 2010. "Policy Statement—Children as Hematopoietic Stem Cell Donors." *Pediatrics* 125:392–404.

——. 2010. "Vaccine Studies—Examine the Evidence." http://www.aap.org/en-us/advocacy-and-policy/Documents/vaccinestudies.pdf.

American College of Obstetricians and Gynecologists. 2008. "Committee Opinion Number 399: On Umbilical Cord Blood Banking." *Obstetrics and Gynecology* 111:475–477.

Anderlik, M. R., and M. A. Rothstein. 2003. "Canavan Decision Favors Researchers Over Families." *Journal of Law and Medical Ethics* 31:450–454.

Andrews, Lori B. 2002. "Genes and Patent Policy: Rethinking Intellectual Property Rights." *Nature Reviews Genetics* 3:803–808.

——. 2005. "Harnessing the Benefits of Biobanks." *Law, Medicine, and Ethics* 33(1): 22–30.

——. 2006. "Who Owns Your Body? A Patient's Perspective on *Washington University v. Catalona." Journal of Law, Medicine, and Ethics* 34:389–407.

Angell, Marcia. 2005. *The Truth About the Drug Companies: How They Deceive Us and What to Do About It.* New York: Random House.

Angevaren, M., G. Aufdemkampe, H. J. J. Verhaar, A. Aleman, and L. Vanhees. 2008. "Physical Activity and Enhanced Fitness to Improve Cognitive Function in Older People Without Known Cognitive Impairment." Cochrane Database of Systematic Reviews, Issue 3. Art. no. CD005381. DOI: 10.1002/14651858.CD005381. pub3.

Angrist, Misha. 2010. *Here Is a Human Being: At the Dawn of Personal Genomics.* New York: Harper Collins.

Annas, George L. 1999. "Waste and Longing: The Legal Status of Placental Blood Banking." *New England Journal of Medicine* 340:1521–1524.

——. 2010. *Worst Case Bioethics: Death, Disaster, and Public Health.* New York: Oxford University Press.

Aronowitz, Robert. 2010. "Gardasil: A Vaccine Against Cancer and a Drug to Reduce Risk." In *Three Shots at Prevention: The HPV Vaccine and the Politics of Medicine's Simple Solutions,* ed. Keith Wailoo et al., 21–38. Baltimore, Md.: Johns Hopkins University Press.

Arranz, M.J., and J. de Leon. 2007. "Pharmacogenetics and Pharmacogenomics of Schizophrenia: A Review of the Last Decade of Research." *Molecular Psychiatry* 12:707–747.

Arribas-Ayllon, Michael. 2010. "Beyond Pessimism: The Dialogue of Promise and Complexity in Genomic Research." *Genomics, Society, and Policy* 6:1–12.

Association of Molecular Pathology et al. v. United States Patent and Trade Office and Myriad Genetics Inc. 2010. 669 F Supp 2d 365.

Ayme, Segolene, Gert Matthijs, and S. Soini. 2008. "Patenting and Licensing in Genetic Testing: Recommendations of the European Society of Human Genetics." *European Journal of Human Genetics* 16: 405–411.

Bainbridge, M.N., et al. 2011. "Whole-Genome Sequencing for Optimized Patient Management." *Science and Translational Medicine* 3(87). DOI: 10.1126/scitranslmed .3002243.

Ballen, Karen. 2010. "Challenges in Umbilical Cord Blood Stem Cell Banking for Stem Cell Reviews and Reports." *Stem Cell Reviews and Reports* 6:8–14.

Barker, J. N., D. Weisdorf, T. E. Defoe, et al. 2005. "Transplantation of 2 Partially HLA-Matched Umbilical Cord Blood Units to Enhance Engraftment in Adults with Hematologic Malignancy." *Blood* 105:1343–1347.

Bates, Benjamin R., et al. 2004. "Evaluating Direct-to-Consumer Marketing of Race-

Based Pharmacogenmics: A Focus Group Study of Public Understandings of Applied Genomic Medicine." *Journal of Health Communication* 9:541.

Battin, Margaret P., Leslie P. Francis, Jay A. Jacobson, and Charles B. Smith. 2008. *The Patient as Victim and Vector: Ethics and Infectious Disease*. New York: Oxford University Press.

Baylis, Francoise. 2009. "For Love or Money? The Saga of the Korean Women Who Provided Eggs for Embryonic Stem Cell Research." *Theoretical Medicine and Bioethics* 30:385–396.

Baylis, Francoise, and Jason Scott Robert. 2006. "Radical Rupture: Exploring Biological Sequelae of Volitional Inheritable Genetic Modification." In *The Ethics of Inheritable Genetic Modification*, ed. John Rasko et al., 131–148. Cambridge: Cambridge University Press.

Beach, W.A., D. W. Easter, J. S. Good, and E. Pigeron. 2005. "Disclosing and Responding to Cancer 'Fears' During Oncology Interviews." *Social Science and Medicine* 60:893–910.

Beatty, P. G., M. Mori, and E. Milford. 1995. "Impact of Racial Genetic Polymorphism Upon the Probability of Finding an HLA-Matched Donor." *Transplantation* 60(8):778–783.

Beauchamp, Thomas, and James Childress. 2009. *Principles of Medical Ethics*. 6th ed. New York: Oxford University Press.

Benet, Stephen Vincent. 1927. *John Brown's Body*. New York: Farrar and Rinehart.

Berry, John M. 2011. "Who Died from Flu?" *New York Review of Books* (June 23).

Bess, Michael. 2011. "Blurring the Boundary Between 'Person' and 'Product.' 2: Human Genetic Engineering Through the Year 2060." *Hedgehog Review* 13(2).

Bewley, Susan. 2011. Personal communication (November 28).

Bienvault, Pierre. 2010. "Sang de cordon: des banques familiales gratuites." *La Croix* (May 25).

Black, C., et al. 2002. "Relation of Childhood Gastrointestinal Disorders to Autism: Nested Case Control Study Using Data from the UK General Practice Research Database." *British Medical Journal* 325:419–421.

Bloss, C. S., N. J. Schork, and E. J. Topol. 2012. "Effect of Genomewide DTC Profiling to Assess Disease Risk." *New England Journal of Medicine*, prepublication release.

Blume, Stuart. 2006. "Antivaccination Movements and Their Implications." *Social Science and Medicine* 62:628–642.

Boggio, Andrea. 2005. "Charitable Trusts and Human Research Genetic Databases: The Way Forward?" *Genomics, Society, and Policy* 2:41–49.

Bolouri, Hamid. 2010. *Personal Genomics and Personalized Medicine*. London: Imperial College Press.

Boot, B. P., B. Partridge, and W. Hall. 2010. "Better Evidence for Safety and Efficacy Is Needed Before Neurologists Prescribe Drugs for Neuroenhancement to Healthy People." *Neurocase* (October 18), e-publication ahead of publication.

Borry, Pascal, Martina C. Cornel, and Heidi C. Howard. 2010. "Where Are You Going, Where Have You Been: A Recent History of the Direct-to-Consumer Genetic Testing Market." *Journal of Community Genetics* 1:101–106.

Boseley, Sarah. 2010. "Wakefield Struck off Medical Register Following Scare Over MMR." *The Guardian* (May 25).

Bostrom, Nick, and Anders Sandberg. 2009. "Cognitive Enhancement: Methods, Ethics, Regulatory Challenges." *Science and Engineering Ethics* 15:311–341.

Boyle, James. 1996. *Shamans, Software, and Spleens: Law and the Construction of the Information Society.* Cambridge, Mass.: Harvard University Press.

——. 2003. "The Second Enclosure Movement and the Construction of the Public Domain." *Law and Contemporary Problems* 66:33–74.

Braithwaite, R. B. 1963. *The Theory of Games as a Tool for the Moral Philosopher.* Cambridge: Cambridge University Press.

Brassington, Iain. 2010. "Enhancing Evolution and *Enhancing Evolution.*" *Bioethics* 24(8):395–401.

Braun, Lundy, and Ling Phoun. 2010. "HPV Vaccination Campaigns: Masking Uncertainty, Losing Complexity." In *Three Shots at Prevention: The HPV Vaccine and the Politics of Medicine's Simple Solutions*, ed. Keith Wailoo et al., 39–60. Baltimore, Md.: Johns Hopkins University Press.

Brockes, Emma. 2012. "Home Truths (Interview with Toni Morrison)." *Guardian Weekend* (April 14).

Brody, Howard. 2009. *The Future of Bioethics.* New York: Oxford University Press.

Brooks, Jamie D., and Meredith L. King. 2008. *Geneticizing Disease: Implications for Racial Health Disparities.* Washington, D.C.: Center for American Progress.

Brooks, M. Alison, and Beth A. Tarini. 2011. "Genetic Testing and Youth Sports." *Journal of the American Medical Association* 305(10):1033–1034.

Brotherton, Julia M. L., Masha Fridman, Cathryn L. May, Genevieve Chappell, A. Marion Saville, and Dorota M. Gertig. 2011. "Early Effect of the HPV Vaccination Programme on Cervical Abnormalities in Victoria, Australia: An Ecological Study." *Lancet* 377:2085–2092.

Brown, Nik. 2011. "The Immunitary Bioeconomy: The Economisation of Life in the International Cord Blood Market." Paper given at the Institute of Advanced Studies /World Universities Network Symposium on Biocapital and Bioequity, University of Bristol, April 27.

Brown, Nik, Laura Machin, and Danae McLeod. 2011. "The Immunitary Bioeconomy: The Economisation of Life in the International Cord Blood Market." *Social Science and Medicine* 30:1–8.

Brown, Phyllida. 2006. "Do We Even Need Eggs?" *Nature* 439:933–637.

Brownsword, Roger. 2009. "Regulating Human Enhancement: Things Can Only Get Better?" *Law, Innovation, and Technology* 1:125–152.

Buchanan, Allen. 2011. *Beyond Humanity? The Ethics of Biomedical Enhancement.* Oxford: Oxford University Press.

Budzyn, D., et al. 2010. "Lack of Association Between Measles-Mumps-Rubella Vaccination and Autism: A Case-Control Study." *Pediatric Infectious Disease Journal* 29(5). http://www.ncbi.nhm.nih.gov/pubmed/19952979.

Burnam-Fink, Michael. 2011. "The Rise and Decline of Military Human Enhancement." *Science Progress* (January 7). http://www.geneticsandsociety.rsvpl.com/article.php?id=5540.

Busby, Helen. 2010. "The Meanings of Consent to the Donation of Cord Blood Stem Cells: Perspectives from an Interview-Based Study of a Public Cord Blood Bank in England." *Clinical Ethics* 5(1):22–27.

Cakic, V. 2009. "Smart Drugs for Cognitive Enhancement: Ethical and Pragmatic Considerations in the Era of Cosmetic Neurology." *Journal of Medical Ethics* 35:611–615.

Calabresi, Guido, and Phillip Bobbitt. 1978. *Tragic Choices.* New York: Norton.

Calderon-Margalit, Ronit, et al. 2009. "Cancer Risk After Exposure to Treatments for Ovulation Induction." *American Journal of Epidemiology* 169(3):365–375.

Caldwell, J. J., J. Caldwell, N. R. Smythe, et al. 2000. "A Double-Blind, Placebo-Controlled Investigation of the Efficacy of Modafinil for Sustaining the Alertness and Performance of Aviators: A Helicopter Simulator Study." *Psychopharmacology* 150:272–282.

Callahan, Daniel. 1984. "Autonomy: A Moral Good, Not a Moral Obsession." *Hastings Center Report* 14(5):4042.

——. 2003. "Individual Good and Common Good: A Communitarian Approach to Bioethics." *Perspectives in Biology and Medicine* 46(4):496–507.

Callaway, Ewen. 2012. "Norway to Bring Cancer-Gene Tests to the Clinic: A Pilot Programme Will See Latest Tumour-Sequencing Techniques to Help Guide Cancer Care." *Nature* (February 2). http://www.nature.com/news/norway-to-bring-cancer-tests-to-the-clinic-1.9949.

Campbell, Denis. 2012. "Healthcare Comes to the Community." *The Guardian* (April 11).

Caplan, Art. 2011. "Did the Vaccine Industry Manipulate the WHO to sell H1N1 Shots?" *Science Progress* (January 20). http://scienceprogress.org/2011/01/did-the -vaccine-industry-manipulate-the-who-to-sell-h1n1-shots/.

Carey, Nessa. 2011. *The Epigenetics Revolution: How Modern Biology Is Rewriting Our Understanding of Genetics, Disease, and Inheritance.* London: Icon.

Carson, Peter, et al. 1999. "Racial Differences in Response to Therapy for Heart Failure: Analysis of the Vasodilator Heart Failure Trial." *Journal of Heart Failure* 5:178–182.

Carvel, John. 2005. "With Love at Christmas: A Set of Stem Cells." *The Guardian* (December 6).

Centers for Disease Control. 2012. "2012 Recommended Immunizations for Children from Birth Through 6 Years Old." http://www.cdc.gov/vaccines/parents/downloads /parent-ver-sch-0-6yrs.pdf.

——. "2012 Recommended Immunizations for Children from 7 Through 18 Years Old." http://www.cdc.gov/vaccines/who/teens/downloads/parent-version-schedule -7-18yrs.pdf.

Chan, Sarah, and John Harris. 2008. "The Ethics of Gene Therapy." *Current Opinion in Molecular Therapeutics* 8:377–383.

Chapman, A. R., and M. S. Frankel, eds. 2003. *Designing Our Descendants: The Promises and Perils of Genetic Modifications.* Baltimore, Md.: Johns Hopkins University Press.

Chaparro, C., L. Neufeld, G. Ten Alavez, et al. 2006. "Effect of Timing of Umbilical Cord Clamping on Iron Status in Mexican Infants: A Randomised Controlled Trial." *Lancet* 367:1997–2004.

Chatterjee, Anjan. 2004. "Cosmetic Neurology: The Controversy Over Moment, Mentation, and Mood." *Neurology* 63:968–974.

Chatterton, Chris. 2010. "Cord Blood Used to Treat Infant with SCID." *Bionews* (November 15). http://www.bionews.org.uk/page_82400.asp?dinfo=z1aVqoC6k5e6EY ReBcoCAIzz.

Chen, Robert T., and Frank DeStefano. 1998. "Vaccine Adverse Events: Causal or Coincidental?" *Lancet* 351(9103):611–612.

Chiang, Alex, and Ryan P. Milton. 2011. "Personalized Medicine in Oncology: Next Generation." *Nature Reviews Drug Discovery* 10:895–896.

Christopher v. Pharmastem Therapeutics, Inc. 2008. NY Westchester Co. Sup. No. 222379/05, March 14.

Clarke, Juanne N., and Michelle M. Everest. 2006. "Cancer in the Mass Print Media: Fear, Uncertainty, and the Medical Model." *Social Science and Medicine* 62(10): 2591–2600.

Coghlan, Andy. 2011. "Is There Life for Stem Cells After Geron?" *New Scientist* (November 18). http://www.newscientist.com/article/dn21190-is-there-life-for-stem-cells-after-geron.html?full=true.

Cohen, Deborah, and Philip Carter. 2010. "WHO and the Pandemic Flu 'Conspiracies.'" *British Medical Journal* 340:1274–1276.

Cohen, Jon. 2012. "WHO Meeting of Flu Experts Calls for Full Publication of Controversial H5N1 Papers." *Science Insider* (February 17). http://news.sciencemag.org/scienceinsider/2012/02/who-meeting-of-flu-experts-calls.html.

Cohn, Vicki. 2008. "Umbilical Cord Blood Stem Cell Therapy in an Animal Model of Alzheimer's Disease." *Genetic Engineering News* (March 26).

Colgrove, James K. 2006a. *State of Immunity: The Politics of Vaccination in Twentieth-Century America*. Berkeley: University of California Press.

——. 2006b. "The Ethics and Politics of Compulsory HPV Vaccination." *New England Journal of Medicine* 355:2389–2391.

——. 2010. "The Coercive Hand, the Beneficent Hand: What the History of Compulsory Vaccination Can Tell Us About HPV Vaccine Mandates." In *Three Shots at Prevention: The HPV Vaccine and the Politics of Medicine's Simple Solutions*, ed. Keith Wailoo et al., 3–20. Baltimore, Md.: Johns Hopkins University Press.

Collins, Francis S. 2010. *The Language of Life: DNA and the Revolution in Personalized Medicine*. New York: Harper Collins.

Collins, Francis S., and V. A. McKusick. 2001. "Implications of the HGP for Medical Science." *Journal of the American Medical Association* 285:540–544.

Comité Central National d'Éthique (CCNE). 2002. "Opinion Number 74: Umbilical Cord Blood Banks for Autologous Use or for Profit." Paris: CCNE.

——. 2012. "Avis no. 117: Utilisation des cellules souches issues du sang de cordon ombilical, du cordon lui-même et du placenta et leur conservation en biobanques. Questionnement ethique." Paris: CCNE.

Condit, Celeste, et al. 2003. "Attitudinal Barriers to Delivery of Race-Targeted Pharmacogenomics Among Informed Lay Persons." *Genetics in Medicine* 5:385.

Cook, Michael. 2010. "The Scourge of Ageing Is Worse Than Smallpox, Says UK Researcher." *Bioedge* (May 14). http://www.bioedge.org/index.php/bioethics/bioethics_article/8998.

Cook-Deegan, Robert, and Christopher Heaney. 2010. "Patents in Genomics and Human Genetics." *Annual Review of Genomics and Human Genetics* 11:383–425.

Cooper, Melinda. 2007. *Life as Surplus: Biotechnics and the Transformation of Capital*. Seattle: University of Washington Press.

Copelan, N., D. Harris, and M. A. Gaballa. 2009. "Human Umbilical Cord Blood Stem Cells, Myocardial Infarction and Stroke." *Clinical Medicine* 9:342–345.

Cotter, A., A. Ness, and J. Tolosa. 2001. "Prophylactic Oxytocin for the Third Stage of Labour." *Cochrane Database of Systematic Reviews* 4.

Council of Europe. 1997. Convention for the Protection of Human Rights and Dignity of the Human Being with Regard to the Application of Biology and Medicine (Convention on Human Rights and Biomedicine, or "Oviedo Convention"). http://conventions.coe.int/Treaty/en/Treaties/html/164.htm.

Cranage, Allison. 2011. "NHS Stem Cell Services get £4 Million Cash Injection." *Bionews* (August 1). http://www.bionews.org.uk/page_102684.asp?dinfo=z1aVqoC6k5e6EY ReBcoCAIzz.

Crockett, Molly. 2009. "Values, Empathy, and Fairness Across Social Barriers." *Annals of the New York Academy of Sciences* 1167:76–86.

Cudkowicz, Merit. 2010. "A Futility Study of Minocycline in Huntington's Disease." *Movement Disorders* 25(13):2219–2224. DOI: 10.1002/mds.23236. PMID 20721920.

Cunningham, Michael. 1999. *The Hours*. New York: Farrar, Straus and Giroux.

Curtis, Christina, Sohrab P. Shah, Suet-Feung Chin, et al. 2012. "The Genomic and Transcriptomic Architecture of 2,000 Tumours Reveals Novel Subgroups." *Nature* (April 18). DOI: 10.1038/nature10983.

Daly, Herman, and John Cobb Jr. 1989. *For the Common Good: Redirecting the Economy Toward Community, the Environment, and a Sustainable Future*. Boston: Beacon.

Daniels, Norman. 2008. *Just Health: Meeting Health Needs Fairly*. Cambridge: Cambridge University Press.

Darnovsky, Marcy. 2008. "The 'Spitterati' and Trickle-Down Genomics." *Biopolitical Times*. http://www.geneticsandsociety.rsvp1.com/article.php?id=4360&mgh=http %3A%2F%2Fwww.geneticsandsociety.org&mgf=1.

——. 2010. "'Moral Questions of an Altogether Different Kind': Progressive Politics in the Biotech Age." *Harvard Law and Policy Review* 4:99–119.

Darwin, Erasmus. 1801. *Zoonomia*. Vol. 3. 3rd ed. London.

Davey, Sue, Sue Armitage, Vanderson Rocha, et al. 2004. "The London Cord Blood Bank: Analysis of Banking and Transplantation Outcome." *British Journal of Haematology* 125(3):358–365.

Davies, Kevin. 2010. *The Thousand-Dollar Genome: The Revolution in DNA Sequencing and the New Era of Personalized Medicine*. New York: Free Press.

Davies, Melanie. 2009. "Assisted Conception: Uses and Abuses." In *Reproductive Ageing*, ed. Susan Bewley, William Ledger, and Dimitrios Nikolaou, 227–236. London: Royal College of Obstetricians and Gynaecologists Press.

Decker, Susan, and Kayla Bruun. 2012. "Myriad Defends Patent Claims on Genetic Material in Court Case." *Bloomberg Businessweek* (July 20). http://www.businessweek

.com/news/2012-07-20/myriad-defends-patent-claims-on-genetic-material-in-court-case#p1.

Deer, Brian. 2004. "MMR Scare Doctor Planning Rival Vaccine." *Times* (London) (November 14). http://www.briander.com/wakefield.patent.htm (referencing patent application number 9711663.6, UK Patent Office, Department of Trade and Industry).

Delden, J. J. van, R. Ashcroft, A. Dawson, G. Marckmann, R. Upshur, and M. F. Verweij. 2008. "The Ethics of Mandatory Vaccination Against Influenza for Health Care Workers." *Vaccine* 28:5562–5566.

DeStefano, F., et al. 2004. "Age of First Measles-Mumps-Rubella Vaccination in Children with Autism and School-Matched Control Subjects: A Population-Based Study in Metropolitan Atlanta." *Pediatrics* 113(2):239–266.

Devine, Karen. 2010. "Private Cord Blood Collection and Cerebral Palsy—Is There a Connection?" *Bionews* (November 22).

Dickenson, Donna. 1997. *Property, Women, and Politics: Subjects or Objects?* Cambridge: Polity.

——. 2003a. *Risk and Luck in Medical Ethics.* Cambridge: Polity.

——. 2003b. "Consent, Commodification, and Benefit Sharing in Genetic Research." *University of Nairobi Law Journal* 1:146–157.

——. 2006. "The Lady Vanishes: What's Missing from the Stem Cell Debate." *Journal of Bioethical Inquiry* 3:43–54.

——. 2008. *Body Shopping: The Economy Fuelled by Flesh and Blood.* Oxford: Oneworld.

——. 2009. "An Uneasy Case Against Stephen Munzer: Umbilical Cord Blood and Property in the Body." *American Philosophical Association Newsletter* 8(2):11–16.

——. 2010. "Dangers of Personal Genetic Testing." *Toronto Globe and Mail* (June).

——. 2011. "Regulating (or Not) Reproductive Medicine: An Alternative to Letting the Market Decide." *Indian Journal of Medical Ethics* 8(3):175–179.

——. 2012a. *Bioethics: All That Matters.* London: Hodder Education.

——. 2013. "Exploitation and Choice in the Global Egg Trade: Emotive Terminology or Necessary Critique?" In *Regulating Contestable Commodities in the Global Body Market: Altruism's Limits,* ed. Michele Goodwin. New York: Cambridge University Press.

Dickenson, Donna, Richard Huxtable, and Michael Parker. 2010. *The Cambridge Medical Ethics Workbook.* 2nd ed. Cambridge: Cambridge University Press.

Dickenson, Donna, and Paolo Vineis. 2002. "Evidence-Based Medicine and Quality of Care." *Health Care Analysis* 10: 243–259.

Dion-Labrie, Marianne, et al. 2010. "The Use of Personalized Medicine for Patient Selection for Renal Transplantation: Physicians' Views on the Clinical and Ethical Implications." *BMS Medical Ethics* 11:5.

Dodds, Susan. 2003. "Women, Commodification, and Embryonic Stem Cell Research." In *Biomedical Ethics Reviews: Stem Cell Research*, ed. James Humber and Robert F. Almeder, 149–175. Totowa, N.J.: Humana.

Doja, A. W., and W. Roberts. 2006. "Immunizations and Autism: A Review of the Literature." *Canadian Journal of Neurological Sciences* 33(4):341–346.

Dondorp, Wybo, and Guido de Wert. 2010. "Monitoring Report: Genetics and Health." The Hague: Center for Ethics and Health.

Doshi, Peter. 2007. "Estimation of Death Rates from Pandemic Influenza." *Lancet* 369(9563): 739.

——. 2008. "Trends in Recorded Influenza Mortality: United States, 1900–2004." *American Journal of Public Health* 98(5):939–945.

Douglas, Mary. 1966. *Purity and Danger: An Analysis of Concepts of Pollution and Taboo*. London: Routledge.

Douglas, Thomas. 2008. "Moral Enhancement." *Journal of Applied Philosophy* 25(3): 228–245.

Downey, Candice L., and Susan Bewley. 2009. "Third Stage Practices and the Neonate." *Fetal and Maternal Medicine Review* 20(3):1–18.

Duncan, David Ewing. 2009. *Experimental Man: What One Man's Body Reveals About His Future, Your Health, and Our Toxic World*. Hoboken, N.J.: Wiley.

Dutfield, Graham. 2009. *Intellectual Property Rights and the Life Science Industries: Past, Present, and Future*. 2nd ed. Singapore: World Scientific.

Dworkin, Gerald. 1988. *The Theory and Practice of Autonomy*. Cambridge: Cambridge University Press.

Ecker, Jeffrey L., and Michael F. Greene. 2005. "Editorial: The Case Against Private Umbilical Cord Blood Banking." *Obstetrics and Gynecology* 105(6):1281–1284.

Ediezen, Leroy C. 2006. "NHS Maternity Units Should Not Encourage Commercial Banking of Umbilical Cord Blood." *British Medical Journal* 333:801–804.

Elliott, Carl. 1998. "The Tyranny of Happiness." In *Enhancing Human Traits: Ethical and Social Implications*, ed. Erik Parens, 177–188. Washington, D.C.: Georgetown University Press.

——. 2004. *Better Than Well: American Medicine Meets the American Dream*. New York: Norton.

Eng, Charis, et al. 2010. "Comparison of Family Health History to Personal Genomic Screening: Which Method Is More Effective for Risk Assessment of Breast, Colon, and Prostate Cancer?" Presented at the American Society of Human Genetics 60th Annual Meeting, November 2–6, Washington, D.C.

Engelhardt, H. Tristram. 1996. *Germline Genetic Engineering and Moral Diversity: Moral Controversies in a Post-Christian World*. New York: Cambridge University Press.

Epstein, Helen. 2011. "Flu Warning—Beware the Drug Companies!" *New York Review of Books* (May 13).

Epstein, Steven. 2010. "The Great Undiscussable—Anal Cancer, HPV, and Gay Men's Health." In *Three Shots at Prevention: The HPV Vaccine and the Politics of Medicine's Simple Solutions,* ed. Keith Wailoo et al., 61–90. Baltimore, Md.: Johns Hopkins University Press.

Epstein, Steven, and April N. Huff. 2010. "Sex, Science, and the Politics of Biomedicine." In *Three Shots at Prevention: The HPV Vaccine and the Politics of Medicine's Simple Solutions,* ed. Keith Wailoo et al., 213–228. Baltimore, Md.: Johns Hopkins University Press.

Etzioni, Amitai. 2011. "Authoritarian Versus Responsive Communitarian Bioethics." *Journal of Medical Ethics* 37(1):17–23.

European Parliament Science and Technology Assessment Office (STOA). 2009. *Human Enhancement.* Brussels: European Commission.

Evans, J. H. 2002. *Playing God? Human Genetic Engineering and the Rationalization of Public Bioethical Debate, 1959–1995.* Chicago: University of Chicago Press.

Farah, Martha J., Judy Illes, Robert Cook-Deegan, et al. 2004. "Neurocognitive Enhancement: What Can We Do and What Should We Do?" *Nature Reviews Neuroscience* 5:421–425.

Fathimathas, Lux. 2010. "Personalised Cancer Therapy on the NHS." *Bionews* (June 5). http://www.bionews.org.uk/page_62605.asp.

Fauser, Bart, et al. 2010. "Mild Ovarian Stimulation for IVF: 10 Years Later." *Human Reproduction* 25(11):2678–2684.

Fenton, Elizabeth. 2010. "The Perils of Failing to Enhance: A Response to Persson and Savulescu." *Journal of Medical Ethics* 36(3):148–151.

Ferreria, E., J. Pasternak, N. Bacal, et al. 1999. "Autologous Cord Blood Transplantation." *Bone Marrow Transplantation* 24:1041.

Fidanboylu, Mehmet. 2011. "US National Institutes of Health Put $416 Million Into Personalised Medicine." *Bionews* 637 (December 12). http://www.bionews.org.uk/page_115244.asp&dinfo=zlaVqoC6k5e6EydBcoCAIzz.

Fidler, David P., and Lawrence O. Gostin. 2011. "WHO's Pandemic Influenza Preparedness Framework: A Milestone in Global Governance for Health." *Journal of the American Medical Association* 306(2):200–201.

Fields, Stanley, and Mark Johnson. 2010. *Genetic Twists of Fate.* Cambridge, Mass.: MIT Press.

Fisk, Nicholas M., and Rifat Atun. 2008. "Public-Private Partnership in Cord Blood Banking." *BMJ* 336:642–644.

Fleck, Leonard M. 2010. "Personalized Medicine's Ragged Edge." *Hastings Center Report* 40(5):16–18.

Flegel, Ken. 2009. "Editorial: Ten Reasons to Make Cord Blood Stem Cells a Public Good." *Canadian Medical Association Journal* 180(13):1279.

Fleming, N. 2008. "Rival Genetic Tests Leave Buyers Confused." *Sunday Times* (September 7).

Fombonne, Eric, et al. 2006. "Pervasive Developmental Disorders in Montreal, Quebec, Canada: Prevalence and Links with Immunizations." *Pediatrics* 118(1):e139–e150. DOI: 10.1542/peds.2005-2993.

Foster, M., J. Mulvihill, and R. Sharp. 2009. "Evaluating the Utility of Personal Genomic Information." *Genetic Medicine* 11:570–574.

Fouchier, Ron A. M., Adolfo Garcia-Sastre, Yoshihiro Kawaoka, et al. 2012. "Pause on Avian Flu Transmission Research" (letter). *Science* (January 20). DOI: 10.1126/science/1219412.

Fowler, James H., and Christopher T. Dawes. 2007. "Two Genes Predict Voter Turnout." *Journal of Politics* 70(3):579–594.

Fox, N., C. Stevens, R. Ciobotariu, et al. 2007. "Cord Blood Collection: Do Patients Really Understand?" *Journal of Perinatal Medicine* 35:314–321.

Fox, Renee C., and Judith P. Swazey. 2008. *Observing Bioethics*. New York: Oxford University Press.

Frank, Lone. 2011. *My Beautiful Genome: Exploring Our Genetic Future, One Quirk at a Time*. Oxford: Oneworld.

Frank, Thomas. 2012a. *Pity the Billionaire: The Hard-Times Swindle and the Unlikely Comeback of the Right*. London: Harvill Secker.

——. 2012b. "We Don't Need No Regulation." *Guardian Weekend* (January 7).

Frankfurt, Harry. 1971. "Freedom of the Will and the Concept of a Person." *Journal of Philosophy* 68:5–20.

Fukuyama, Francis. 2002. *Our Posthuman Future: Consequences of the Biotechnology Revolution*. London: Profile.

GAO (Government Accountability Office). 2010. "Direct-to-Consumer Genetic Tests: Misleading Test Results Are Further Complicated by Deceptive Marketing and Other Questionable Practices." Highlights of GAO-10–847T, testimony before Subcommittee on Oversight and Investigations, Committee on Energy and Commerce, House of Representatives, July 22.

Garland, Suzanne M., Julia M. Brotherton, John R. Condon, et al. 2011. "Human Papillomavirus Prevalence Among Indigenous and Nonindigenous Australian Women Prior to a National HPV Vaccination Program." *BMC Medicine* 9:104–117.

Geddes, Linda. 2011. "Daily Aspirin Cuts Risk of Colorectal Cancer." *New Scientist* (October 28).

Gènéthique. 2011a. "France: changement de politique sur le sang de cordon." March 3. http://www.genethique.org/revues/revues/2011/Mars/20110303.1.asp.

——. 2011b. "Le conseil d'État note sa décision." May 6. http://www.genethique.org /revues/revues/2011/Mai/20110506.1.asp.

——. 2011c. "Conservation de tissues et de cellules: vers une interdiction des banques privées?" November 30. http://www.genethique.org/revues/revues/2011/Novembre /20111130.1.asp.

Genewatch UK. 2003. *Pharmacogenetics: Better, Safer Medicines?* Buxton: Genewatch.

——. 2010. *History of the Human Genome.* Buxton: Genewatch.

Genomics Law Report. 2011. "GLR: Genetic Bill of Rights Proposed in Massachusetts." February 14. http://www.genomicslawreport.com/index.php/2011/02/14 /genetic-bill-of-rights-proposed-in-massachusetts.

Gerlinger, Marco, Andrew J. Rowan, Stuart Horswell, et al. 2012. "Intratumor Heterogeneity and Branched Evolution Revealed by Multiregion Sequencing." *New England Journal of Medicine* 366(10):883–892.

George, Rose. 2012. "Dirty Little Secret." *Guardian Weekend* (February 4).

Gilbert, Susan. 2009. "Prescrbing Cognitive Enhancers: A Primer." Bioethics Forum, October 28. www.thehastingscenter.org.

——. 2010. "Personalized Cancer Care in an Age of Anxiety." *Hastings Center Report* 40(5):18–21.

Gillon, Raanan. 1986. *Philosophical Medical Ethics.* Chichester: Wiley.

Ginsburg, Geoffrey S., and F. Willard Huntington. 2009. "Genomic and Personalized Medicine: Foundations and Applications." *Translational Research* 154(6):277–287.

Glannon, Walter. 2007. "Cognitive Memory and Memory Enhancement." Bioethics Forum, May 21. www.thehastingscenter.org/bioethicsforum.

Gluckman, Eliane. 2001. "Hematopoietic Stem-Cell Transplants Using Umbilical-Cord Blood." *New England Journal of Medicine* 344:1860–1861.

Gluckman, Eliane, H. A. Broxmeyer, A. D. Auerbach, et al. 1989. "Hematopoietic Reconstitution in a Patient with Fanconi's Anemia by Means of Umbilical Cord Blood from an HLA-Identical Sibling." *New England Journal of Medicine* 321:1174–1178.

Goetz, Thomas. 2010. *The Decision Tree: Taking Control of Your Health in the Era of Personalized Medicine.* Emmaus, Penn.: Rodale.

Goldacre, Ben. 2008. *Bad Science.* New York: Harper.

Goldman, Brian. 2007. "HER2: The Patent 'genee' Is out of the Bottle." *Journal of the Canadian Medical Association* 176:14433–11444.

Goldstein, David B., and Gianperro L. Cavalleri. 2005. "Understanding Human Diversity." *Nature* 437:1241.

Goldstein, David B., Sarah K. Tate, and Sanjay M. Sisodiya. 2003. "Pharmacogenetics Goes Genomic." *Nature Reviews Genetics* 4:937.

Goodwin, Michele. 2006. *Black Markets: The Supply and Demand of Body Parts.* New York: Cambridge University Press.

Gorner, Peter. 2000. "Parents Suing Over Patenting of Genetic Test." *Chicago Tribune* (November 19).

Gottlieb, Karen. 1998. "Human Biological Samples and the Law of Property: The Trust as a Model for Biological Repositories." In *Stored Tissue Samples: Ethical, Legal, and Public Policy Implications,* ed. R. F. Weir, 183–197. Iowa City: University of Iowa Press.

Gould, Benjamin. 2009. "Cognitive Enhancement on Campus: Taking Competition Seriously." Bioethics Forum, January 21. www.thehastingscenter.org/bioethicsforum.

Greek National Bioethics Commission. 2007. *Opinion on Cord Blood Banking.* Athens: Greek National Bioethics Commission.

Greeley, Henry T., Barbara Sahakian, John Harris, et al. 2008. "Toward Responsible Use of Cognitive-Enhancing Drugs by the Healthy." *Nature* 702.

Greenberg et al. v. Miami Children's Hospital Research Institute et al. 2003. 264 F. Supp 2nd 1064.

Grimley Evans, J., R. Malouf, F. A. H. Huppert, and J. K. Van Niekerk. 2006. "Dehydroepiandrosterone (DHEA) Supplementation for Cognitive Function in Healthy Elderly People." Cochrane Database of Systematic Reviews 2006, Issue 4. Art. No.: CD006221. DOI: 10.1002/14651858.CD006221.

Gruber, Jeremy. 2011. "Personalized Medicine: The Promise of the Genomic Revolution." Presentation at the Working Session on Genetic Testing and Personalized Medicine, the Tarrytown Meetings (Center for Genetics and Society), Tarrytown, N.Y., July 26.

Gundle, Kenneth R., Molly J. Dingle, and Barbara A. Koenig. 2010. "'To Prove This Is the Industry's Best Hope': Big Tobacco's Support of Research on the Genetics of Tobacco Addiction." *Addiction* 105:974–983.

Gura, Trisha. 2012. "Genomics, Plain and Simple: A Pennsylvania Clinic Working with Amish and Mennonite Communities Could Be a Model for Personalized Medicine." *Nature* 283:20–22.

Habermas, Jürgen. 2003. *The Future of Human Nature.* Cambridge: Polity.

Haley, Rebecca, Liana Horvath, and Jeremy Sugarman. 1997. "Ethical Issues in Cord Blood Banking: Summary of a Workshop." *Transfusion* 38:367–373.

Hall, S. S. 2003. "The Quest for a Smart Pill." *Scientific American* (September): 54–65.

Hall, Stuart. 2011. "The March of the Neoliberals." *Guardian* (September 13).

Hall, W., P. Madden, and N. Linskey. 2002. "The Genetics of Tobacco Use: Methods, Findings, and Policy Implications." *Tobacco Control* 11:119–124.

Haller, J. M., H. L. Viener, C. Wasserfall, et al. 2008. "Autologous Umbilical Cord Blood Infusion for Type 1 Diabetes." *Experimental Hematology* 36:710–715.

Hamburg, Margaret A., and Francis S. Collins. 2010. "The Path to Personalized Medicine." *New England Journal of Medicine* 363:301–304. DOI: 10.1056/NEJM1006304.

Hardin, Garrett. 1968. "The Tragedy of the Commons." *Science* 162:1243.

Harmon, Amy. 2010. "Indian Tribe Wins Fight to Limit Research of Its DNA," *New York Times* (April 22).

Harper, Matthew. 2011. "Cancer's New Era of Promise and Chaos." *Forbes* (June 5). http://www.forbes.com/sites/matthewherper/2011/06/05/cancers-new-era-of -promise-and-chaos/.

Harris, Anna, Sally Wyatt, and Susan Kelly. 2012. "The Gift of Spit (and the Obligation to Return It): How Consumers of Online Genetic Testing Services Participate in Research." *Information, Communication, and Society.* DOI: 10.1080/1369118X.2012.701656.

Harris, James W. 1996. *Property and Justice.* Oxford: Clarendon.

Harris, J. M., J. Mauer, and A. L. Kellerman. 2010. "Influenza Vaccine—Safe, Effective, and Mistrusted." *New England Journal of Medicine* 363:2183–2185.

Harris, John. 1985. *The Value of Life.* London: Routledge.

——. 2007. *Enhancing Evolution.* Princeton, N.J.: Princeton University Press.

——. 2010. "God on Moral Enhancement." *Bioethics* 25(2):1–16.

Harris, John, and Soren Holm. 1995. "Is There a Moral Obligation Not to Infect Others?" *BMJ* 311:1215–1217.

Harrosh, Shlomit. 2012. "Moral Enhancement and the Duty to Eliminate Evildoing." Paper presented at the Moral Evil in Practical Ethics conference, University of Oxford, January 21.

Hastings Center blog. 2011. April 4.

Hathaway, Feighanne, Esther Burns, and Harry Ostrer. 2007. "Consumers' Desire Toward Current and Prospective Reproductive Genetic Testing." *Journal of Genetic Counseling* 18(2):137–146.

Hatzis, Christos, et al. 2011. "A Genomic Predictor of Response and Survival Following Taxane-Anthracycline Chemotherapy for Invasive Breast Cancer." *Journal of the American Medical Association* 305(18):1873–1881.

Haug, Charlotte J. 2008. "Human Papillomavirus Vaccination—Reasons for Caution." *New England Journal of Medicine* 359(8):861–862.

Hayden, Cori. 2007. "Taking as Giving: Bioscience, Exchange, and the Politics of Benefit Sharing." *Social Studies of Science* 37:729–760.

Hebert, Daniel W., Zhang Ge, and Elliot S. Vessell. 2008. "From Human Genetics and Genomics to Pharmacogenetics and Pharmacogenomics: Past Lessons, Future Directions." *Drug Metabolism Review* 40(2):187–224.

Hedgecoe, Adam. 2004. "Critical Bioethics: Beyond the Social Science Critique of Applied Ethics." *Bioethics* 18(2):120–143.

——. 2007. *The Politics of Personalised Medicine: Pharmacogenetics in the Clinic*. Cambridge: Cambridge University Press.

Hedges, Chris. 2010. *Death of the Liberal Class*. New York: Nation Books.

Heger, Monica. 2010. "UK Hospital to Sequence the Exomes of 10,000 Heart Disease Patients on SOLiD," www.genomeweb.com/sequencing/UK-hospital-sequence-exomes-10000-heart-disease-patients-solid.

Heller, Michael A. 1998. "The Tragedy of the Anticommons: Property in the Transition from Marx to Markets." *Harvard Law Review* 111:621–688.

Hermitte, Marie-Angèle. 1996. *Le sang et le droit: essai sur la transfusion sanguine*. Paris: Éditions du Seuil.

Heron, J., and J. Golding. 2004. "Thimerosal Exposure in Infants and Developmental Disorders: A Prospective Cohort Study in the United Kingdom Does Not Support a Causal Association." *Pediatrics* 114:577–583.

Hoffman, Jan. 2008. "Vaccinating Boys for Girls' Sake?" *New York Times* (February 24).

Hogarth, Stuart. 2009. "Direct-to-Consumer Genetic Testing: Regulatory Issues." Conference paper, Ethox Centre, University of Oxford, May 21.

Hogarth, Stuart, and Paul Martin. 2012. "The Myth of the Genomic Revolution." *Bionews* 643 (February 6). http://www.bionews.org.uk/page_123189.asp&dinfo=z1aVqoC6k5e6EYReBcoCAIzz.

Honoré, A. M. 1987. "Ownership." In *Making Law Bind: Essays Legal and Philosophical*, 161–192. Oxford: Clarendon.

Hoppe, Nils. 2009. *Bioequity—Property and the Human Body*. Burlington, Vt.: Ashgate.

Hornig, M., et al. 2008. "Lack of Association Between Measles Virus Vaccine and Autism with Enteropathy: A Case-Control Study." *PloS One* 3(9):e3140. DOI: 10.1371/journal.pone.0003140.

Hudspeth, Robert, ed. 1993. *The Letters of Margaret Fuller*. Vol. 6. Ithaca, N.Y.: Cornell University Press.

Huffington, Arianna. 2002. "Finding the Answer to Washington's Hottest Whodunit." *Huffington Post* (December 4).

HUGO (Human Genome Organization). 1996. "Summary of Principles Agreed at the International Strategy Meeting on Human Genome Sequencing ('Bermuda Statement')." London: Wellcome Trust.

HUGO (Human Genome Organization) Ethics Committee. 2000. "Statement on Benefit-Sharing." Geneva: World Health Organization. http://www.hugo-international.org/hugo/benefit.html.

Human Genetics Commission (UK). 2010. "A Common Framework of Principles for Direct-to-Consumer Genetic Testing Services." London: HGC. http://www.hgc .gov.uk/Client/document.asp?DocId=280&CAtegoryId=10.

Hutchon, David J. R. 2008. "Rapid Response: Cord Clamping and Cord Blood Banking." *BMJ* 336 (March 22).

Hutton, E. K., and E. S. Hassan. 2007. "Late vs. Early Clamping of the Umbilical Cord in Full-Term Neonates: Systematic Review and Meta-Analysis of Controlled Trials." *Journal of the American Medical Association* 297(11):1241–1252.

Huxley, Aldous. 1994. *Brave New World*. New York: HarperCollins.

Hviid, A., M. Stellfield, J. Wohlfahrt, and M. Melbye. 2003. "Association Between Thimerosal-Containing Vaccine and Autism." *Journal of the American Medical Association* 290:1763–1766.

Hwang, Woo-Suk. 2004. "Evidence of a Pluripotent Human Stem Cell Line Derived from a Cloned Blastocyst." *Science* 303:1669–1674.

——. 2005. "Patient-Specific Embryonic Stem Cells from Human SCNT Blastocysts." *Science* 308:1777–1783.

Ikemoto, Lisa. 2009. "Eggs as Capital: Human Egg Procurement in the Fertility Industry and the Stem Cell Research Enterprise." *Signs* 4:763–781.

Illich, Ivan. 1995. *Deschooling Society: The Essential Argument Against Schooling*. 2nd ed. New York: Marion Boyars.

Institute of Medicine of the National Academies. 2012. "Evolution of Translational Omics: Lessons Learned and the Path Forward." March 23. http://www.iom.edu/~/ media/Files/Report%20Files/2012/Translational-Omics/omics_rb.pdf.

Jacobs, Allen, James Dwyer, and Peter Lee. 2001. "Seventy Ova." *Hastings Center Report* 31:12–14.

Jacobson, Robert M., Paul V. I. Targonski, and Gregory A. Poland. 2007. "A Taxonomy of Reasoning Flaws in the Antivaccine Movement." *Vaccine* 25:3146–3152.

Jayakrishnan, T. 2011. "Newer Vaccines in the Universal Immunisation Programme." *Indian Journal of Medical Ethics* 8(2):107–112.

Jefferson, T., M. Jones, P. Doshi, and C. Del Mar, C. 2009. "Neuraminidase Inhibitors for Preventing and Treating Influenza in Healthy Adults: Systematic Review and Meta-analysis." *BMJ* 339:b5106.

Jegede, Ayodele Samuel. 2007. "What Led to the Public Boycott of the Nigerian Vaccination Campaign?" *PloS Medicine* 4: e73. DOI: 10.1371/journal/pmed.0040073.

Jensen, K., and F. Murray. 2005. "International Patenting: The Landscape of the Human Genome." *Science* 310:239–240.

Jesani, Amar. 2011. "Policy and Global Health: How Can Things Change?" Paper given at the Wellcome Trust conference on global public health, London, June 22.

John, T. Jacob, and N. K. Shah. 2011. "Editorial: Universal Healthcare and Nationwide Public Health: A Tale of Two Declarations from One City." *Indian Journal of Medical Research* 134:250–252.

Johnson, L. Syd M. 2010. "Lessons from H1N1 (Swine Flu)." Bioethics Forum, Hastings Center, February 3. http://www.thehastingscenter.org/Bioethicsforum/Post.aspx?id=4454&blogid=140.

Johnson, Linda A. 2012. "GlaxoSmithKline Wins Takeover of Human Genome Sciences." *Washington Post* (July 16). http://www.huffingtonpost.com/2012/07/16/glaxosmithkline-human-genome_n_1676019.html.

Johnson, Summer. 2010. "Ethics of Reproductive Tourism Questioned." *Blog.Bioethics .Net* (May 20). http://www.bioethics.net/tags/blog-bioethics-net.

Jonas, Hans. 1982. *The Imperative of Responsibility: In Search of an Ethics for the Technological Age*. Chicago: University of Chicago Press.

Kahn, Jonathan. 2004. "How a Drug Becomes 'Ethnic': Law, Commerce, and the Production of Racial Categories in Medicine." *Yale Journal of Health Policy, Law, and Ethics* 4(1):1–16.

——. 2007. "Race in a Bottle." *Scientific American* (August): 41–45.

——. 2011. "Translational Budgets." *Biopolitical Times* (February 2).

Kahneman, D., P. Slovic, and A. Tversky, eds. 1982. *Judgment Under Uncertainty*. New York: Cambridge University Press.

Kaimal, Anjali, et al. 2009. "Cost-Effectiveness of Private Umbilical Cord Blood Banking." *Obstetrics and Gynecology* 114(4):848–855. DOI: 10.1097/AOG .0b013e3181b8fcod.

Kalf, R. R. J., R. Mihaescu, P. de Knijff, et al. 2011. "Risk Predictions from Direct-to-Consumer Personal Genome Testing: What Do Consumers Really Learn About Common Disease Risk?" Paper presented at the European Society for Human Research and Embryology (ESHRE) conference, Amsterdam, May.

Kant, Immanuel. 1949. *Fundamental Principles of the Metaphysic of Morals*. Trans. Thomas K. Abbott. Indianapolis, Ind.: Bobbs-Merrill.

Kimberg, D. Y., M. Esposito, and M. J. Farah. 1997. "Effects of Bromocriptine on Human Subjects Depend on Working Memory Capacity." *Neuroreport* 8:3581–3585.

Kimmelman, Jonathan. 2010. *Gene Therapy and the Ethics of First-in-Humans Research: Lost in Translation*. Cambridge: Cambridge University Press.

Kirschenbaum, Sheila R. 1997. "Banking on Discord: Legal Conflicts in the Transplantation of Umbilical Cord Blood Cells." *Arizona Law Review* 39:1391–1396.

Kitching, Aileen, et al. 2010. "Oseltamivir Side-Effects and Adherence Among Children in Three London Schools Affected by Influenza A(H1N1)v—An Internet-Based Cross-Sectional Survey." Poster abstract presented at the "Understanding

Behavioural Repsonses to Infectious Disease Outbreaks" conference, Regents College, London, June 11.

Klein, Naomi. 2008. *The Shock Doctrine: The Rise of Disaster Capitalism*. London: Penguin.

Knafo, A. et al. 2008. "Individual Differences in Allocation of Funds in the Dictator Game Associated with Length of the Arginine Vasopressin 1a Receptor RS3 Promoter Region and Correlation Between RS3 Length and Hippocampal mRNA." *Genes, Brain, and Behaviour* 7:266–275.

Knoppers, Bartha M. 2005. "Biobanking: International Norms." *Journal of Law, Medicine, and Ethics* 33(1):7–14.

Knoppers, Bartha M., Denise Avard, and Heidi C. Howard. 2010. "Direct-to-Consumer Genetic Testing: Driving Choice?" *Expert Reviews in Molecular Diagnostics* 10(8): 965–968.

Kolata, Gina. 2011. "How the Bright Promise in Cancer Testing Fell Apart." *New York Times* (July 7).

Kollman, C. 2004. "Assessment of Optimal Size and Composition of the U.S. National Registry of Hematopoietic Stem Cell Donors." *Transplantation* 78(1):89–93.

Kramer, W., J. Schneider, and N. Schultz. 2009. "U.S. Oocyte Donors: A Retrospective Study of Medical and Social Issues." *Human Reproduction* (September 3).

Krimsky, Sheldon. 2011. "A Neoliberal Economics of Science." *American Scientist* 99(4):330.

Kuczynski, Alex. 2006. *Beauty Junkies: Inside Our $15 Billion Obsession with Cosmetic Surgery*. New York: Doubleday.

Kwak, E. L., et al. 2010. "Anaplastic Lyphoma Kinase Inhibition in Non-Small-Cell Lung Cancer." *New England Journal of Medicine* 363:1695–1703.

Laberge, Anne-Marie, and Wylie Burke. 2008. "Personalized Medicine and Genomics." In *From Birth to Death and Bench to Clinic: The Hastings Center Bioethics Book for Journalists, Policymakers, and Campaigns*, ed. Mary Crowley, 133–136. Garrison, N.Y.: Hastings Center.

Lamkin, Matt. 2011. "Regulating Enhancements: Preserving Normal Functioning by Reducing Its Significance." Paper presented at the Tarrytown Meetings (Center for Genetics and Society), July 25.

Lancet. 2010. "New Guidelines for Genetic Tests Are Welcome but Insufficient" (editorial). *Lancet* 376:488.

Larriviere, Dan, et al., on behalf of the American Academy of Neurology Ethics, Law, and Humanities Committee. 2009. "Responding to Requests from Adult Patients for Neuroenhancements: Guidance of the Ethics, Law, and Humanities Committee." *Neurology* 73(17):1406–1412.

Larson, Heidi J., Pauline Brocard, and Geoffrey Garnett. 2010. "The India HPV Vaccine Suspension." *Lancet* 376(9741):572–573.

Laskey, L. C., et al. 2002. "In Utero or Ex Utero Cord Blood Collection: Which Is Better?" *Transfusion* 42: 1261–1267.

Le Brocq, Michael. 2008. Quoted in Virgin Health Bank press release (April 18), "Patients at Risk from Critical Shortfall in Cord Blood Stem Cells."

Le Doeuff, Michele. 2000. *Le sexe du savoir.* Paris: Champs Flammarion.

Lee, Soo-Youn, and Howard L. McLeod. 2011. "Pharmacogenetic Tests in Cancer Chemotherapy: What Physicians Should Know for Clinical Application." *Journal of Pathology* 223(1):15–27.

Lees, K. A., P. M. Wortley, and S. S. Coughlin. 2005. "Comparison of Racial/Ethnic Disparities in Adult Immunization and Cancer Screening." *American Journal of Preventive Medicine* 29(5):404–411.

Leigh, Bertie. 2005. "Umbilical Cord Blood Stem Cell Banking—Legal Review." London: Royal College of Obstetricians and Gynaecologists.

Lethaby, A., E. Hogervorst, M. Richards, et al. 2008. "Hormone Replacement Therapy for Cognitive Function in Postmenopausal Women." *Cochrane Database of Systematic Reviews* 1. Art. no.: CD003122. DOI: 10.1002/14651858.CD003122.pub2.

Levin, Mark. 2012. Speech at Eighth Annual State of Personalized Medicine Luncheon, May 8, National Press Club, Washington, D.C.

Levitt, Mairi, and Sue Weldon. 2005. "A Well Placed Trust? Public Perceptions of the Governance of DNA Databases." *Critical Public Health* 15(4):311–321.

Lim, W. S., J. K. Gammack, J. K. Van Niekerk, and A. Dangour. 2006. "Omega 3 Fatty Acid for the Prevention of Dementia." *Cochrane Database of Systematic Reviews* 1. Art. no.: CD005379. DOI: 10.1002/14651858.CD005379.pub2.

Ling, Tom, and Ann Raven. 2005. "Pharmacogenetics and Uncertainty: Implications for Policy Makers." *Studies in History and Philosophy of Biological and Biomedical Sciences* 37(3):533–549.

Lingam, R., A. Simmons, N. Andrews, et al. 2003. "Prevalence of Autism and Parentally Reported Triggers in a North East London Population." *Archives of Disease in Childhood* 88(6):666–670.

Link, Daniel C., et al. 2011. "Identification of a Novel TP53 Cancer Susceptibility Mutation Through Whole-Genome Sequencing of a Patient with Therapy-Related AML." *Journal of the American Medical Association* 305(15):1568–1576.

Linn, Sabine. 2012. "When to Add Chemotherapy to Endocrine Therapy and Endocrine Sensitivity." Abstract presented at the Eighth European Breast Cancer Conference, Vienna, March 22.

Lippman, Abby. 1998. "The Politics of Health: Geneticization Versus Health Promotion." In *The Politics of Health: Exploring Agency and Autonomy*, ed. Susan Sherwin, 64–82. Philadelphia: Temple University Press.

Livingston, Julie, Keith Wailoo, and Barbara M. Cooper. 2010. "Vaccination as Governance: HPV Skepticism in the United States and Africa, and the North-South Divide." In *Three Shots at Prevention: The HPV Vaccine and the Politics of Medicine's Simple Solutions*, ed. Keith Wailoo et al., 231–253. Baltimore, Md.: Johns Hopkins University Press.

Lo, M. D., et al. 2010. "Maternal Plasma DNA Sequencing Reveals the Genetic and Mutational Profile of the Fetus." *Science and Translational Medicine* 2:61.

Locatelli, F., V. Rocha, W. Reed, et al. (Eurocord Transplant Group) 2003. "Related Umbilical Cord Blood Transplantation in Patients with Thalassemia and Sickle Cell Disease." *Blood* 101:2137–2143.

Lorizio, Wendy, Hugo Rugo, Mary S. Beattie, et al. 2011. "Pharmacogenetic Testing Affects Choice of Therapy Among Women Considering Tamoxifen Treatment." *Genome Medicine* 3:64.

Lynch, G. 2002. "Memory Enhancement: The Search for Mechanism-Based Drugs." *Nature Neuroscience* 5:1035–1038.

Lyons, Rachel. 2012. "Personalised Care One Step Closer as Mayo Clinic Sequences Patients' Genomes." *Bionews* (January 12). http://www.bionews.org.uk/page_116341 .asp&dinfo=z1aVqoC6k5e6EYReBcoCAIzz.

Ma, N., C. Stamm, A. Kaminski, et al. 2005. "Human Cord Blood Cells Induce Angiogenesis Following Myocardial Infarction in NOD/scid Mice." *Cardiovascular Research* 66:45–54.

MacFarlane, Jo. 2009. "Swine Flu Jab Link to Killer Nerve Disease: Leaked Letter Reveals Concern of Neurologists Over 25 Deaths in America." *Daily Mail* (London) (August 15). http://www.dailymail.co.uk/news/article-1206807.

Madsen, K. M., M. B. Lauritsen, C. B. Pedersen, et al. 2003. "Thimerasol and the Occurrence of Autism: Negative Ecological Evidence from Danish Population-Based Data." *Pediatrics* 112:604–606.

Madsen, K. M., et al. 2002. "A Population-Based Study of Measles, Mumps, and Rubella Vaccination and Autism." *New England Journal of Medicine* 347(19):1477–1482.

Maher, Brendan. 2011. "Human Genetics: Genomes on Prescription." *Nature* 478:22–24.

Malak, Janet, and Judith Daar. 2012. "The Case for a Parental Duty to use Preimplantation Genetic Diagnosis for Medical Benefit." *American Journal of Bioethics* 12(4): 3–11.

Malakoff, David, and David Enserink. 2012. "In Dramatic Move, Researchers Announce Moratorium on Some H5N1 Flu Research, Call for Global Summit." *Science Insider* (January 20).

Malouf, R., and J. Grimley Evans. 2008. "Folic Acid with or Without Vitamin B12 for the Prevention and Treatment of Healthy Elderly and Demented People." *Cochrane Database of Systematic Reviews* 4. Art. no.: CD004514. DOI: 10.1002/14651858. CD004514.pub2.

Martin, M., L. Clare, A. M. Altgassen, et al. 2012. "Cognition-Based Interventions For Healthy Older People and People with Mild Cognitive Impairment." *Cochrane Database of Systematic Reviews* 1 (2011). Art. no.: CD006220. DOI: 10.1002/14651858 .CD006220.pub2.

Maxmen, Amy. 2011. "Pharmacogenomics: Playing the Odds." *Nature* 474:S9–S10. DOI: 10.1038/474S9a.

Mayo Collaborative Services, DBA Mayo Medical Laboratories et al. v. Prometheus Laboratories. No. 10-1150, Supreme Court of the United States, 132 S. Ct. 1289; 182 L. Ed. 2d 321; 2012 U.S. LEXIS 2316; 80 U.S.L.W. 4225; 101 U.S.P.Q.2D (BNA) 1961, December 7, 2011, argued; March 20, 2012, decided.

Mazzucato, Mariana. 2012. "Without State Funds There'd Be No Google or Glaxo." *Guardian* (April 23).

McCabe, Sean Esteban, John R. Knight, Christian J. Teter, and Henry Wechsler. 2005. "Nonmedical Use of Prescription Stimulants Among U.S. College Students: Prevalence and Correlates from a National Survey." *Addiction* 99:96–106.

McConnell, Terrance. 2010. "Genetic Intervention and the Parent-Child Relationship." *Genomics, Society, and Policy* 6(3):1–14.

McGowan, Michelle L., Jennifer R. Fishman, and Marcie A. Lambrix. 2010. "Personal Genetics and Individual Identities: Motivations and Imperatives of Early Users." *New Genetics and Society* 29(3):261–290.

McGuire, A. L., et al.. 2009. "Social Networkers' Attitudes Towards DTC Genetic Testing," *American Journal of Bioethics* 9(6–7):3–10.

McKenna, David, and Jayesh Sheth. 2011. "Umbilical Cord Blood: Current Status and Promise for the Future." *Indian Journal of Medical Research* 134:261–269.

McLeod, Carolyn, and Francoise Baylis. 2006. "Feminists on the Inalienability of Human Embryos." *Hypatia* 21(1):1–24.

Meghani, Zehra. 2010. "A Robust, Particularist Ethical Assessment of Medical Tourism." *Developing World Bioethics* 11:16–29.

Mehlman, M. J., and J. R. Botkin. 1998. *Access to the Genome: The Challenge of Equality*. Washington, D.C.: Georgetown University Press.

Mercer, J. S. 2001. "Current Best Evidence: A Review of the Literature on Umbilical Cord Clamping." *Journal of Midwifery and Women's Health* 46: 402–414.

Merz, Jon F., et al. 2002. "Protecting Subjects' Interest in Genetic Research." *American Journal of Human Genetics* 70(4):965–971.

Meyer, Emily Ann, Kathi Hanna, and Christine Gebbie, eds. 2004. *Cord Blood: Establishing a National Hematopoietic Stem Cell Bank Program.* Washington, D.C.: Institute of Medicine of the National Academies.

Mirowski, Philip. 2011. *Science-Mart: Privatizing American Science.* Cambridge, Mass.: Harvard University Press.

Mitchell, John. 1907. "A Prayer." *Life* 1268 (February 14).

Mnookin, Seth. 2011. *The Panic Virus: A True Story of Medicine, Science, and Fear.* New York: Simon and Schuster.

Moise, Kenneth J., Jr. 2005. "Umbilical Cord Stem Cells." *Obstetrics and Gynecology* 106:1393–1407.

Moore v. Regents of the University of California. 1990. 793 P2d 479.

Moreno, J. D. 2006. *Mind Wars: Brain Research and National Defense.* New York: Dana.

Morley, Rosie. 2011. "UK Cancer Charity Launches Project to Develop Personalized Medicines." *Bionews* (November 28). http://www.bionews.org.uk/page.asp?obj_id=113141&print=1.

Morrison, Toni. 2012. *Home.* New York: Knopf.

Mukherjee, Siddhartha. 2011. *The Emperor of All Maladies.* New York: Fourth Estate.

Munzer, Stephen D. 1999. "The Special Case of Property Rights in Umbilical Cord Blood for Transplantation." *Rutgers Law Review* 51:493–568.

Munzer, Stephen D., and Franklin O. Smith. 2001. "Limited Property Rights in Umbilical Cord Blood for Transplantation and Research." *Journal of Pediatric Hematology /Oncology* 23(4):203–207.

Murch, Simon, et al. 2004. "Retraction of an Interpretation." *Lancet* 363(9411):750.

Murphy, J. 2004. "Distrust of U.S. Foils Effort to Stop Crippling Disease." *Baltimore Sun* (January 4).

Murray, Christopher J. L., et al. 2007. "Estimation of Potential Global Pandemic Influenza Mortality on the Basis of Vital Registry Data from the 1918–1920 Pandemic: A Quantitative Analysis." *Lancet* 368:2211–2217.

Mykitiuk, Roxanne. 2000. "The New Genetics in the Post-Keynesian State." In *The Gender of Genetic Futures: The Canadian Biotechnology Strategy: Assessing Its Effects on Women and Health*, ed. F. Miller, L. Weir, R. Mykitiuk, et al. Proceedings of the National Strategic Workshop on Women and the New Genetics. Toronto: NNEWH Working Paper Series.

National Research Council (U.S.). 2008. *Emerging Cognitive Neuroscience and Related Technologies*. Washington, D.C.: National Academies Press.

——. 2009. *Opportunities in Neuroscience for Future Army Applications*.

Nature Biotechnology. 2005. "Editorial." *Nature Biotechnology* 23:903.

Neary, Marianne. 2010. "Genetic Link to Fertility Drug Response Found." *Bionews* 565 (July 5). http://www.bionews.org.uk/page_65069.asp?dinfo=z1aVqoC6k5e6EY ReBcoCAIzz.

Nebert, Daniel W., Zhang Ge, and E. S. Vessell. 2008. "From Human Genetics and Genomics to Pharmacogenetics and Pharmacogenomics: Past Lessons, Future Directions." *Drug Metabolism Review* 40(2):187–224.

Nelkin, Dorothy, and M. Susan Lindee. 1995. *The DNA Mystique: The Gene as Cultural Icon*. New York: W. H. Freeman.

New York Times (editorial). 2012. "An Engineered Doomsday." *New York Times* (January 7).

Newman, Stuart A. 2010. "The Transhumanism Bubble." *Capitalism, Nature, Socialism* 21(2):29–42.

Nicoll, Angus. 2010. "The Pandemic from an ECDC (European Centre for Disease Prevention and Control) Perspective." Keynote address given at workshop on "Lessons Learned from Swine Flu for Behavioural and Social Scientists: How Should We Study the Next Pandemic?" Wellcome Trust Conference Centre, London, April 12.

Nordgren, Anders. 2010. "Personal Genomics, Consumer Genomics Companies, and Their Rhetoric." In *Consumer Medicine*, ed. Aaro Tupasela. Copenhagen: Nordic Council of Ministers.

Nuffield Council on Bioethics. 2007. *Public Health: Ethical Issues*. London: Nuffield Council on Bioethics.

——. 2010. *Medical Profiling and Online Medicine: The Ethics of "Personalised Healthcare" in a Consumer Age*. London: Nuffield Council on Bioethics.

——. 2012. "What Are Novel Neurotechnologies?" Webpage for Working Party on Novel Neurotechnologies. http://www.nuffieldbioethics.org/neurotechnology/neuro technology-what-are-novel-neurotechnologies?.

Obasogie, Osagie. 2009. *Playing the Gene Card? A Report on Race and Human Biotechnology*. Oakland, Calif.: Center for Genetics and Society.

Office of Science and Technology Policy. 2012. "National Bioeconomy Blueprint Released." http://www.whitehouse.gov/blog/2012/04/26/national-bioeconomy-blue print-released.

Offit, Paul A. 2007. *Autism's False Prophets: Bad Science, Risky Medicine, and the Search for a Cure*. New York: Columbia University Press.

O'Neill, Onora. 2002. *Autonomy and Trust in Bioethics*. Cambridge: Cambridge University Press.

Orkin, S. H., and A. G. Motulsky. 1995. *Report and Recommendations of the Panel to Assess the NIH Investment in Research on Gene Therapy*. National Institutes of Health.

Ostrom, Elinor. 1990. *Governing the Commons: The Evolution of Institutions for Collective Action*. Cambridge: Cambridge University Press.

Otten, J., H. Wyle, and G. Phelps. 2004. "The Charitable Trust as a Model for Genomic Banks." *New England Journal of Medicine* 350:85–86.

Pacificord. 2010. http://www.pacificord.com/1_0_Stem-Cells-Family-Cord-Blood-Bank .php.

Paik, Young-Gyung. 2007. Personal communication, June 11.

Palsson, Gisli. 2009. "Spitting Image." *Anthropology Now* 1(3):1–22.

Parens, Erik. 2005. "Authenticity and Ambivalence: Toward Understanding the Enhancement Debate." *Hastings Center Report* 35(3):35–41.

Parker, Michael, ed. 1999. *Ethics and Community in the Health Care Professions*. London: Routledge.

———. 2012. *Ethical Problems and Genetics Practice*. Cambridge: Cambridge University Press.

Parliament of India (Rajya Sabha), Department-Related Parliamentary Standing Committee on Health and Family Welfare. 2012. *Fifty-Ninth Report on the Functioning of the Central Drugs Standard Control Organisation (CDSCO)*. New Delhi: Rajya Sabha Secretariat.

Partridge, B. J., et al. 2011. "'Smart Drugs as Common as Coffee': Media Hype About Neuroenhancement." *PLoS One* 6(11):e28416.

Pateman, Carole, and Charles W. Mills. 2007. *Contract and Domination*. Cambridge: Polity.

Paxman, Rosemary. 2010. "Wait Before Cutting the Umbilical Cord, Say Researchers." *Bionews* 560 (May 29).

Paynter, Nina P., et al. 2010. "Association Between a Literature-Based Genetic Risk Score and Cardiovascular Events in Women." *Journal of the American Medical Association* 303(7):631–637.

Peltola, H., et al. 1998. "No Evidence for Measles, Mumps, and Rubella Vaccine-Associated Inflammatory Bowel Disease or Autism in a 14-Year Prospective Study." *Lancet* 351:1327–1328.

Personalized Medicine Coalition. 2009. *The Case for Personalized Medicine*. May.

Persson, Ingmar, and Julian Savulescu. 2008. "The Perils of Cognitive Enhancement and the Urgent Imperative to Enhance the Moral Character of Humanity." *Journal of Applied Philosophy* 25(3).

PHG Foundation. 2010. "Public Health in an Era of Genome-Based and Personalised Medicine." Cambridge: PHG Foundation.

Pitman, R., K. Sanders, R. Zusman, et al. 2002. "Pilot Study of Secondary Prevention of Posttraumatic Stress Disorder with Propranolol." *Biological Psychiatry* 51:189–192.

Poland, Gregory A., and Robert M. Jacobson. 2011. "The Age-Old Struggle Against the Antivaccinationists." *New England Journal of Medicine* 364:97–99.

Pollack, Andrew. 2010. "Awaiting the Genome Payoff." *New York Times* (June 15).

Porter, Eduardo. 2007. "The Divisions That Tighten the Purse Strings." *New York Times* (April 29).

Potti, Anil, et al. 2006. "Genomic Signatures to Guide the Use of Chemotherapeutics." *Nature Medicine* 12:1294–1300.

Prainsack, Barbara. 2011. *Solidarity.* London: Nuffield Council on Bioethics.

Prasad, V. K., A. Mendizabal, S. H. Parikh, et al. 2008. "Unrelated Donor Umbilical Cord Blood Transplantation for Inherited Metabolic Disorders in 159 Pediatric Patients from a Single Center: Influence of Cellular Composition of the Graft on Transplantation Outcomes." *Blood* 112:2979–2989.

President's Council of Advisors on Science and Technology (PCAST). 2008. *Priorities for Personalized Medicine.* Washington, D.C.: PCAST.

——. 2009. *Report to the President on U.S. Preparations for the 2009 H1N1-Influenza.* Washington, D.C.: PCAST.

President's Council on Bioethics. 2002. *Beyond Therapy.* Washington, D.C.: National Bioethics Advisory Commission.

Price, Christopher S., et al. 2010 "Prenatal and Infant Exposure to Thimerosal from Vaccines and Immunoglobins and Risk of Autism." *Pediatrics* 126(4):656–664. DOI: 10.1542/peds.2010-0309.

Prince, Alfred M. 1977. "Our Last Vaccine?" *Science* 195(4284):1287.

Proulx, S., L. Heaton, J. K. Choon, and M. Millette. 2011. "Paradoxical Empowerment of *Produsers* in the Context of Informational Capitalism." *New Review of Hypermedia and Multimedia* 17(1):9–29.

Putnam, Robert D. 2000. *Bowling Alone: The Collapse and Revival of American Community.* New York: Simon and Schuster.

——. 2008. "The Rebirth of American Civic Life." *Boston Globe* (March 2).

Pywell, S. 2000. "Vaccination and Other Altruistic Medical Treatments: Should Autonomy or Communitarianism Prevail?" *Medical Law International* 4(3–4):223–243.

R. v. Kelly. 1998. 3 All England Reports 741.

Rabe, H., G. Reynolds, and J. Diaz-Rossello. 2004. "Early Versus Delayed Clamping in Preterm Infants." *Cochrane Database Systematic Reviews* 4: CD0032248.

Ramogola-Masire, Doreen. 2010. "Cervical Cancer, HIV, and the HPV Vaccine in Botswana." In *Three Shots at Prevention: The HPV Vaccine and the Politics of Medi-*

cine's Simple Solutions, ed. Keith Wailoo et al., 91–100. Baltimore, Md.: Johns Hopkins University Press.

Rasko, John E. J., and Douglas J. Jolly. 2006. "The Science of Inheritable Genetic Modification." In *The Ethics of Inheritable Genetic Modification*, ed. J. Rasko et al., 17–32. Cambridge: Cambridge University Press.

Rasko, John, Gabrielle O'Sullivan, and Rachel Ankeny, eds. 2006. *The Ethics of Inheritable Genetic Modification*. Cambridge: Cambridge University Press.

Rawls, John. 1971. *A Theory of Justice*. Cambridge, Mass.: Harvard University Press.

Read, Jon. 2010. "Introduction: Defining the Problem." Paper given at the "Understanding Behavioural Responses to Infectious Disease Outbreaks" conference, Regents College, London, June 11.

Reich, Jennifer A. 2010. "Parenting and Prevention: Views of HPV Vaccines Among Parents Challenging Childhood Immunization." In *Three Shots at Prevention: The HPV Vaccine and the Politics of Medicine's Simple Solutions*, ed. Keith Wailoo et al., 165–181. Baltimore, Md.: Johns Hopkins University Press.

Reich, Robert. 2007. *Supercapitalism: The Battle for Democracy in an Age of Big Business*. New York: Knopf.

Richards, Martin. 2010. "Reading the Runes of my Genome: A Personal Exploration of Retail Genetics." *New Genetics and Society* 29:291–310.

Rieff, David. 2008. *Sea of Death: A Son's Memoir*. New York: Simon and Schuster.

Robson, Mark E., et al. 2010. "American Society of Clinical Oncology Policy Statement Update: Genetic and Genomic Testing for Cancer Susceptibility." *Journal of Clinical Oncology* 28(5):893–901.

Roberts, Dorothy. 2010. Preface to Osagie Obasogie, "Playing the Gene Card," *Biopolitical Times*.

——. 2011. *Fatal Invention: How Science, Politics, and Big Business Re-Create Race in the Twenty-First Century*. New York: New Press.

Robertson, John A. 2003. "The $1,000 Genome: Ethical and Legal Issues in Whole-Genome Sequencing of Individuals." *American Journal of Bioethics* 3(3):35–42.

Robson, Mark E., et al. 2008. "American Society of Clinical Oncology Policy Statement Update: Genetic and Genomic Testing for Cancer Susceptibility." *Journal of Clinical Oncology* 28(5):893–901.

Rocha, V. et al. 2000. "Graft-Versus-Host Disease in Children Who Have Received a Cord-Blood or Bone-Marrow Transplant from an HLA-Identical Sibling." *New England Journal of Medicine* 25:1846–1854.

Rocha, V., M. Labopin, G. Sanz, et al. 2004. "Transplants of Umbilical Cord Blood or Bone Marrow from Unrelated Donors in Adults with Acute Leukemia." *New England Journal of Medicine* 351:2276–2285.

Roe, F. J. C. 1988. "Comments on Draft Proposal from Jeffrey Idle, St. Mary's Hospital Medical School and University of Newcastle upon Tyne, April 21." Cited in Wallace (2009). *

Rogers, J., J. Wood, R. McCandlish, et al. 1998. "Active Versus Expectant Management of the Third Stage of Labour: The Hinchingbrooke Randomised Controlled Trial." *Lancet* 351:693–699.

Rogers, Robert. 2012. "Neural Chemical Systems Mediate Social Behaviour in the 'Tragedy of the Commons': Implications for Ethics and the Clinic." Wellcome Lecture in Neuroethics, February 29, University of Oxford.

Rose, Nikolas. 2007. *The Politics of Life Itself: Biomedicine, Power, and Subjectivity in the Twenty-First Century*. Princeton, N.J.: Princeton University Press.

Roussel, Frederic, and Marie Lechner. 2011. "Transhumanistes sans gêne." *Libération* (June 21).

Roxby, Philippa. 2011. "Measles Outbreak Warning as Cases Rise in Europe and UK." *BBC News* (May 14). http://www.bbc.co.uk/news/health-13378119.

Royal College of Obstetricians and Gynaecologists (RCOG) Scientific Advisory Committee. 2001. "Opinion Paper 2: Cord Blood Banking." London: RCOG.

——. 2006. "Umbilical Cord Blood Banking: Scientific Advisory Committee Opinion Paper 2." London: RCOG.

Royal Society (UK). 2012. *Brain Waves Module 3: Neuroscience, Conflict, and Security*. London: Royal Society.

Rubin, G. J. 2010. "General Population Reactions to Major Public Health Incidents." Paper presented at "Understanding Behavioural Responses to Infectious Disease Outbreaks" conference, Regents College, London, June 11.

Rubin, G. J., R. Amlot, L. Page, and S. Wessely. 2009. "Public Perceptions, Anxiety, and Behavioural Change in Response to the Swine Flu Epidemic: A Cross-Sectional Telephone Survey." *BMJ* 339:b2651.

Salter, Brian. 2007. "Patenting, Morality, and Human Stem Cells Science: Bioethics and Cultural Politics in Europe." *Regenerative Medicine* 2(3):301–311.

Salzberg, Steven L., and Mihaela Pertea. 2010. "Do-It-Yourself Genetic Testing." *Genome Biology* 11:404.

Sample, Ian. 2012. "Rise of the Man-Machines: How Troops Could Plug Their Brains Into Weapons." *Guardian* (February 7).

Samuel, Gabrielle. 2010. "Gene Linked to Blood Clots During Cancer Therapy." *Bionews* (June 21). http://www.bionews.org.uk/page_64704.asp?dinfo=z1aVqoC6k5e6EYReBcoCAIzz.

Sandel, Michael. 1984. "The Procedural Republic and the Unencumbered Self." *Political Theory* 12(1):81–96.

———. 2004. "The Case Against Perfection." *The Atlantic.* http://www.theatlantic.com /past/docs/issues/2004/04/sandel.htm.

———. 2007. *The Case Against Perfection: Ethics in the Age of Genetic Engineering.* Cambridge, Mass.: Harvard University Press.

———. 2012. *What Money Can't Buy: The Moral Limits of Markets.* New York: Farrar, Straus and Giroux.

Sanders, David, and Jesse Dukeminier Jr. 1968. "Medical Advance and Legal Lag: Hemodialysis and Kidney Transplantation." *UCLA Law Review* 15:367–368.

Sarojini, N. B., et al. 2010. "The HPV Vaccine: Science, Ethics, and Regulation." *Economic and Political Weekly* 45:27–34.

Sarojini, N. B., et al. 2011. "Undeniable Violations and Unidentifiable Violators." *Economic and Political Weekly* 46:17–19.

Sartre, Jean-Paul. 2003. *Being and Nothingness: An Essay on Phenomenological Ontology.* London: Routledge.

Sasaki, E., H. Suemizu, A. Shimada, et al. 2009. "Generation of Transgenic Nonhuman Primates with Germline Transmission." *Nature* 459:523–527.

Saunders, Ruth. 2012. "U.S. Health Institute Launches Genetic Test Database." *Bionews* (March 5). http://www.bionews.org.uk/page_131508.asp&dinfo=dmPaop4Jruu2M7 KJMedosTid.

Savulescu, Julian. 2001. "Procreative Beneficence: Why We Should Select the Best Children." *Bioethics* 15:413–426.

———. 2002. "Personal Choice: Letter from a Doctor as a Dad." In *Healthcare Ethics and Human Values*, ed. K. W. M. Fulford et al., 109–110. Oxford: Blackwell.

———. 2003. "Is the Sale of Body Parts Wrong?" *Journal of Medical Ethics* 29:138–139.

Sawaya, George F., and Karen Smith-McCune. 2007. "HPV Vaccination: More Answers, More Questions." *New England Journal of Medicine* 356:1991–1993.

Schatten, Gerald, and S. Mitalipov. 2009. "Transgenic Primate Offspring." *Nature* 459:515–516.

Schechter, R., and J. K. Grether. 2008. "Continuing Increases in Autism Reported to California's Developmental Services System: Mercury in Retrograde." *Archives of General Psychiatry* 65(1):19–24.

Schermer, M. 2008. "Enhancement, Easy Shortcuts, and the Richness of Human Activities." *Bioethics* 22:355–363.

Schwartz, Jason L. 2012. "New Media, Old Messages: Themes in the History of Vaccine Hesitancy and Refusal." *Virtual Mentor (American Medical Association)* 14(1):50–55.

Secretary's Advisory Committee on Genetics, Health, and Society. 2010. *Gene Patent and Licensing Practices and Their Impact on Patient Access to Genetic Tests.*

Bethesda, Md.: National Institutes of Health (NIH) Office of Biotechnology Activities (OBA).

Sengupta, Amit, et al. 2011. "Letter: Human Papillomavirus Trials in India." *Lancet* 377:719.

Senituli, Lopeti. 2004. "They Came for Sandalwood, Now the B—s Are After Our Genes!" Paper presented at the "Research Ethics, Tikanga Maori/Indigenous and Protocols for Working with Communities" conference, Wellington, New Zealand, June 10–12.

Sequist, L. V., R. S. Heist, A. T. Shaw, et al. 2011. "Implementing Multiplexed Genotyping of Non-Small-Cell Lung Cancers Into Clinical Practice." *Annals of Oncology* 22(12):2616–2624.

Shanks, Pete. 2011. "Advertising as Threat: Will Your Kid Die from Exercise?" http://www.biopoliticaltimes.org/article.php?id=5743.

Sharp, Daniel. 2012. "This Time It's Personal: New Cystic Fibrosis Drug (with $294,000 Price Tag)." *Biopolitical Times* (May 15). http://www.biopoliticaltimes.org/article.php?id=6207.

Sherwin, Susan. 1992. *No Longer Patient: Feminist Ethics and Health Care.* Philadelphia: Temple University Press.

Shiva, Vandana. 1997. *Biopiracy: The Plunder of Nature and Knowledge.* Boston: South End.

Sidaway v. Board of Governors of the Bethlem Royal Hospital, 1 All ER 1018.

Sigurdsson, Skuli. 2001. "Yin-Yang Genetics, or the HSE deCODE Controversy." *New Genetics and Society* 20:103–117.

Singer, Emily. 2011. "A Family Learns the Secrets of Its Genomes." *Technology News* (September 16). http://www.technologyreview.com/biomedicine/38612.

Sjogren, Ebba. 2010. "Upsetting Categories? The Consequences of Pharmacogenomics for Making Knowledge-Based Reimbursement Decisions in Sweden." *New Genetics and Society* 29(4):389–411.

Skloot, Rebecca. 2010. *The Immortal Life of Henrietta Lacks.* New York: Crown.

Smeeth, L., et al. 2004. "MMR Vaccination and Pervasive Developmental Disorders: A Case-Control Study." *Lancet* 364(9438):963–969.

Smith, Jennifer S. 2008. "Ethnic Disparities in Cervical Cancer Illness Burden and Subsequent Care: A Prospective View." *American Journal of Managed Care* 14:S193–S199.

Smith, M., and C. Woods. 2010. "On-Time Vaccine Receipt in the First Year Does Not Adversely Affect Neuropsychological Outcomes." *Pediatrics* 125(6):1134–1141.

Sosman, J. A., K. B. Kim, L. Schuchter, et al. 2012. "Survival in BRAF V600-Mutant Advanced Melanoma Treated with Vemurafenib." *New England Journal of Medicine* 366(8):707–714.

Sparrow, Andrew. 2011. "NHS for Sale—Labour Sounds Alarm Over Cameron's Vision." *Guardian* (December 5).

Spring, Joel H. 1971. *Education and the Rise of the Corporate State.* Cuernavaca: CIDOC.

Srinivasan, Sandhya. 2011. "HPV Vaccine Trials and Sleeping Watchdogs." *Indian Journal of Medical Ethics* 8(2):73–74.

Stehr-Green, P., P. Tull, M. Stellfield, et al. 2005. "Autism and Thimerasol-Containing Vaccines: Lack of Consistent Evidence for an Association." *American Journal of Preventive Medicine* 25:101–106.

Steinbrook, Robert. 2005. "Gag Clauses in Clinical Trials Agreements." *New England Journal of Medicine* 352:2180–2182.

Sterckx, Sigrid, and Julian Cockbain. 2012. "The Phenotype Goldmine—Direct-to-Consumer Genetic Testing Companies Discover Patents." Paper delivered at the third IAS/WUN Workshop on Biocapital and Bioequity, University of Bristol, February 21.

Stone, Jay. 2011. "Genetic Test May Help Us Predict Treatment Success in Breast Cancer." *Bionews* (May 16). http://www.bionews.org.uk/page_94564.asp.

Sulston, John, and Georgina Ferry. 2003. *The Common Thread: Science, Politics, Ethics, and the Human Genome.* London: Corgi.

Tang, Y.-P., E. Shimizu, G. Dube, et al. 1999. "Genetic Enhancement of Learning and Memory in Mice." *Nature* 401:63–69.

Tang, Y.-P., E. Shimizu, and J. Tsien. 2001. "Do 'Smart' Mice Feel More Pain, or Are They Just Better Learners?" *Nature Neuroscience* 4:453–454.

Tate, Sarah K., and David B. Goldstein. 2004. "Will Tomorrow's Medicines Work for Everyone?" *Nature Genetics* 36:S34.

Ter Meulen, Ruud, and Donna Dickenson. 2002. "Into the Hidden World Behind Evidence-Based Medicine." *Health Care Analysis* 10:1–20.

Terry, Sharon. 2002. Presentation on PXE Foundation. Paper presented at "Commercialisation of Human Genomics: Consequences for Science and Humanity" conference, Duke University. http://www.law.duke.edu/conference/gelp/program.html.

Thompson, William W., et al. 2007. "Early Thimerosal Exposure and Neuropsychological Outcomes at 7 to 10 Years." *New England Journal of Medicine* 357:1281–1292.

Thornhill, Allen. 2008. "Pharmacogenetics in Assisted Reproduction: Optimizing Response to Ovarian Stimulation." *Bionews* (September 25).

Thornley, Ian, Mary Eapen, Lillian Sung, et al. 2009. "Private Cord Blood Banking: Experiences and Views of Pediatric Hematopoietic Cell Transplantation Physicians." *Pediatrics* 123(3):1011–1017. DOI: 10.1542/peds.2008-0436.

Titmuss, Richard. 1971. *The Gift Relationship: From Human Blood to Social Policy.* Ed. Ann Oakley and J. Ashton. London: London School of Economics.

Tolosa, J. N., Dong-Hyun Park, David J. Eve, et al. 2010. "Mankind's First Natural Stem Cell Transplant." *Journal of Cellular and Molecular Medicine* 14(3): 488–495.

Tomljenovic, Lucija, and Christopher A. Shaw. 2011. "Human Papillomavirus (HPV) Vaccine Policy and Evidence-Based Medicine: Are They at Odds?" *Annals of Medicine* (December 22). DOI: 10.3109/07853890.2011.645353.

Tong, Rosemarie, Anne Donchin, and Susan Dodds, eds. 2004. *Linking Visions: Feminist Bioethics, Human Rights, and the Developing World.* Lanham, Md.: Rowman and Littlefield.

Topol, Eric J. 2012. *The Creative Destruction of Medicine: How the Digital Revolution Will Create Better Health Care.* New York: Basic Books.

Towers, S., and Z. Feng. 2009. "Pandemic H1N1 Influenza: Predicting the Course of a Pandemic and Assessing the Efficacy of the Planned Vaccination Programme in the United States." *Euro Surveillance* 14(41):11. http://www.eurosurveillance.org/view/article.aspx?articled+19358.

Tremblay, Marie-Eve, Aurelie Closon, Guy D'Anjou, and Jean-Francois Buissieres. 2010. "Guillain-Barré Syndrome Following H1N1 Immunization in a Pediatric Patient." *Annals of Pharmacotherapy* 44:1330–1333.

Tressell, Robert. 2005. *The Ragged Trousered Philanthropists.* Ed. Peter Miles. Oxford: Oxford World's Classics.

Triggle, Nick. 2012. "PIP Breast Implants: MPs Condemn NHS Stance in England." *BBC News* (March 28). http://www.bbc.co.uk/news/health-17528830.

Tromp, Saskia. 2001. "Seize the Day, Seize the Cord." Unpublished undergraduate medical dissertation, University of Maastricht.

Tully, T., et al. 2003. "Targeting the CREB Pathway for Memory Enhancers." *Nature Reviews Drug Discovery* 2:267–277.

Tupasela, Aaro. 2010. *Introduction to Consumer Medicine.* Copenhagen: Nordic Council of Ministers.

Turner, Stephen J., Peter C. Doherty, and Anne Kelso. 2010. "Q and A: H1N1 Pandemic Influenza—What's New?" *BMC Biology* 8:130. http://www.biomedcentral.com/1741-7007/8/130.

Tutton, R., and B. Prainsack. 2011. "Entrepreneurial or Altruistic Selves? Making up Research Subjects in Genetic Research." *Sociology of Health and Illness* 33(7):1081–1095.

Twenge, Jean M., and Keith W. Campbell. 2009. *The Narcissism Epidemic: Living in the Age of Entitlement.* New York: Free Press.

Uchiyama, T., et al. 2007. "MMR-Vaccine and Regression in Autism Spectrum Disorders: Negative Results Presented from Japan." *Journal of Autism and Developmental Disorders* 37(2):210–217.

UNESCO. 1997. "Universal Declaration on the Human Genome and Human Rights." http://portal.unesco.org/en/ev.php-URL_ID=13177&URL_DO=DO_TOPIC&URL _SECTION=201.html.

Ungchusak, K., et al. 2005. "Probable Person-to-Person Transmission of Avian Influ- enza A (H5N1)." *New England Journal of Medicine* 352:333–340.

United Health Center for Health Reform and Modernization. 2012. "Personalized Medicine: Trends and Prospects for the New Science of Genetic Testing and Mo- lecular Diagnostics." Working Paper 7 (March).

United Kingdom Health Protection Agency. 2011. "Measles Cases Surpass 2010 Total but MMR Vaccine Uptake Reaches 90 Percent Level for First Time in 13 Years." http://www.hpa.org.uk/NewsCentre/NationalPressReleases/2011PressReleases/ 110624Measlesstatement/.

United States National Academies. 2005. "Report Proposes Structure for National Network of Cord Blood Stem Cell Banks." http://www.sciencedaily.com/releases/ 2005/04/050418095036.htm.

Van Rheenen, P. F., and B. J. Brabin. 2004. "Late Umbilical Cord Clamping as an Inter- vention for Reducing Iron Deficiency Anaemia in Term Infants in Developing and Industrialised Countries: A Systematic Review." *Annals of Tropical Paediatrics* 24:3–16.

Vance, Ashly. 2010. "Merely Human? That's So Yesterday." *New York Times* (June 12).

Vawter, D. 1998. "An Ethical and Policy Framework for the Collection of Umbilical Cord Blood Cells." In *Stored Tissue Samples: Ethical, Legal, and Public Policy*, ed. by R. Weir. Iowa City: Iowa University Press.

Venter, J. Craig et al. 2001. "The Sequence of the Human Genome." *Science* 291: 1304–1351.

Virgin Health Bank. 2011. http://www.virginhealthbank.com/our-services/community -banking.

Wade, Nicholas. 2010. "A Decade Later, Genetic Map Yields Few Cures." *New York Times* (June 12).

Wadman, Meredith. 2011. "Fifty Genome Sequences Reveal Breast Cancer's Complex- ity." *Nature News* (April 2). DOI: 10.1038/news.2011.203.

Wakefield, Andrew. 1998. "Correspondence: Author's Reply: Autism, Inflammatory Bowel Disease, and MMR Vaccine." *Lancet* 351(9106):908.

——. 2012. *Waging the War on the Autistic Child*. New York: Sky Horse.

Wakefield, Andrew, S. H. Murch, A. Anthony, et al. 1998. "Ileal-Lymphoid-Nodular- Hyperplasia: Nonspecific Colitis and Pervasive Developmental Disorder in Chil- dren." *Lancet* 351(9106):637–641.

Waldby, Catherine, and Melinda Cooper. 2010. "From Reproductive Work to Regenera- tive Labour: The Female Body and the Stem Cell Industries." *Feminist Theory* 11:3–22.

Waldby, Catherine, and Robert Mitchell. 2006. *Tissue Economies: Blood, Organs, and Cell Lines in Late Capitalism.* Durham, N.C.: Duke University Press.

Wallace, Helen. 2009. "Big Tobacco and the Human Genome: Driving the Scientific Bandwagon?" *Genomics, Society, and Policy* 5(1):80–133.

——. 2010. Personal communication, July 13.

Walsh, S. Z. 1968. "Maternal Effects of Early and Late Clamping of the Umbilical Cord." *Lancet* 1:996–997.

Warwick, Ruth, and Susan Armitage. 2004. "Best Practice and Research." *Clinical Obstetrics and Gynecology* 18(6):995–2011.

Warren, Oliver J., Daniel R. Leff, Thanos Athanasiou, et al. 2009. "The Neurocognitive Enhancement of Surgeons: An Ethical Perspective." *Journal of Surgical Research* 152(1):167–172.

Washington, Harriet A. 2011. "Flacking for Big Pharma." *American Scholar* (Summer): 22–34.

Washington University v. William J. Catalona, 437 F Supp 2d, ESCD Ed Mo 2006.

Wasserman, David T. 2003. "My Fair Baby: What's Wrong with Parents Genetically Enhancing Their Children?" in *Genetic Prospects: Essays on Biotechnology, Ethics, and Public Policy,* ed. Verna V. Gehring, 99–110. Lanham, Md.: Rowman and Littlefield.

Weeks, Andrew. 2007. "Umbilical Cord Clamping After Birth." *BMJ* 335:312–313.

White, Peter. 2010. "The Epidemiological Modelling Perspective." Paper presented at the "Understanding Behavioural Responses to Infectious Disease Outbreaks" conference, Regents College, London, June 11.

White, Ross. 2011. "Booster(ism) for Personalized Medicine." *Hastings Center Bioethics Forum* (April 4). http://www.thehastingscenter/org/bioethicsforum.

White House. 2012. *National Bioeconomy Blueprint.* Washington, D.C.: The White House.

Widdows, Heather. 2011. "The Poverty of Choice." Inaugural lecture delivered at the University of Birmingham, March 24.

Wiemels, J. L., et al. 1999. "Prenatal Origin of Acute Lymphoblastic Leukaemia in Children." *Lancet* 352:1499–1503.

Wijlaars, Linda. 2012. "'Goldilocks' Gene Response to TB Suggests Best Treatment." *Bionews* (February 6). http://www.bionews.org.uk/page_122899.asp&dinfo=z1aVqoC6k5e6EYReBcoCAIzz.

Wilkie, Andrew. 2009. "Direct-to-Consumer Genetic Testing: Scientific Issues." Paper presented at the "Direct-to-Consumer Genetic Testing: Ethical and Regulatory Issues" conference, Ethox Centre, University of Oxford, May 21.

Wilkinson, Richard, and Kate Pickett. 2010. *The Spirit Level: Why Equality Is Better for Everyone.* Harmondsworth: Penguin.

Winickoff, David E. 2007. "Partnership in UK Biobank: A Third Way for Genomic Governance?" *Journal of Law, Medicine, and Ethics* (Fall): 440ff.

Winickoff, David E., and Larissa B. Neumann. 2005. "Towards a Social Contract for Genomics: Property and the Public in the "Biotrust" Model." *Genomics, Society, and Policy* 1(3):8–21.

Winickoff, David E., and Richard N. Winickoff. 2003. "The Charitable Trust as a Model for Genomic Biobanks." *New England Journal of Medicine* 12:1180–1184.

World Health Organization (WHO). 2002. *Genomics and World Health: Report of the Advisory Committee on Health Research*. Geneva: WHO.

——. 2007. "Recommendations for the Prevention of Postpartum Haemorrhage." Geneva: WHO.

——. 2010. "Programmes and Projects—Global Alert and Response (GAR)—Diseases Covered by GAR-Pandemic (H1N1) 2009—Briefing Notes." http://www.who.int/disease/swineflu/notes/briefing_20091119/en.

Worthey, Elizabeth A., et al. 2011. "Making a Definitive Diagnosis: Successful Clinical Application of Whole Exome Sequencing in a Child with Intractable Inflammatory Bowel Disease." *Genetics in Medicine* 13:255–262.

Wright, Caroline F., and Hilary Burton. 2009. "The Use of Cell-Free Fetal Nucleic Acids in Maternal Blood for Noninvasive Prenatal Diagnosis." *Human Reproduction Update* 15(1):139–151.

Yarwood, Mary. 2012. "MPs Call for More Cord Blood Donations." *Bionews* (February 6). http://www.bionews.org.uk/page_122824.asp&dinfo=z1aVqoC6k5e6EYReBcoCAIzz.

Yearworth v. North Bristol NHS Trust. 2009. 3 Weekly Law Reports 118.

Yesavage, J., M. Mumenthaler, J. Taylor, et al. 2001. "Donepazil and Flight Simulator Performance: Effects on Retention of Complex Skills." *Neurology* 59:123–125.

You, S. H. 2011. "Pharmacoeconomic Evaluation of Warfarin Pharmacogenomics." *Expert Opinion in Pharmacotherapy* (January 14).

Zhao, Yong, Zhaoshun Jiang, Tingbao Zhao, et al. 2012. "Reversal of Type 1 Diabetes Via Islet β Cell Regeneration Following Immune Modulation by Cord Blood-Derived Multipotent Stem Cells." *BMC Medicine* 10:3–12. http://www.biomedcentral.com/1741-7015/10/3.

Zilbertsein, Moshe, Michael Feingold, and Michelle M. Selbel. 1997. "Umbilical Cord-Blood Banking: Lessons Learned from Gamete Donation." *Lancet* 349:642–645.

INDEX